Life in the Universe

Life in the Universe
The Abundance of Extraterrestrial Civilizations

JAMES N. PIERCE

Brown Walker Press
Boca Raton, Florida

Life in the Universe:
The Abundance of Extraterrestrial Civilizations

BrownWalker Press
Boca Raton , Florida
USA • 2008

ISBN-10: 1-59942-451-7 *(pbk)*
ISBN-13: 978-1-59942-451-4 *(pbk)*

ISBN-10: 1-59942-452-5 *(ebook)*
ISBN-13: 978-1-59942-452-1 *(ebook)*

www.brownwalker.com

Library of Congress Cataloging-in-Publication Data

Pierce, James Newsome, 1948-
Life in the universe : the abundance of extraterrestrial civilizations / James N. Pierce.
      p. cm.
Includes bibliographical references and index.
ISBN 1-59942-451-7 (pbk. : alk. paper)
   1. Extraterrestrial beings. 2. Life on other planets. 3. Space environment. 4. Exobiology. 5. Life--Origin. I. Title.

QB54.P54 2008
576.8'39--dc22

2008014079

To Kathy Faber-Langendoen,

Ed Cheng,

Katie Dusenbery,

James Attarian,

and

Carol Penning

Contents

Preface

This is a book about life in the universe, intended for use in introductory college courses on this subject. This topic has generated considerable interest among the general public, especially fans of science-fiction movies and books. Indeed, with the increasing skills of special-effects teams and makeup artists, some of the movies have become quite believable in their appearance – if not in their portrayal of physics – to the degree that many moviegoers accept the existence of extraterrestrial beings as *truth* and their presence here on Earth as *fact*.

Many students entering typical 'Life in the Universe' classes are sure that at least *some* of the tales of encounters with UFOs must be real, that Earth is frequently visited by beings from another world, that Earthlings are occasionally abducted for testing aboard extraterrestrial spaceships, and that spacecraft are currently streaking among the stars to bring more alien scientists eager to study our biology and our culture.

Other students are less easily convinced: they will believe in aliens only when their existence is 'proven' conclusively. They may assert that suitable conditions for life cannot *possibly* exist beyond our own planet, that despite the billions of stars in our galaxy, we are unique.

Which of these viewpoints is closer to the truth? Do aliens exist? Are they aware of *us*? Do they have the means and the desire to travel here to study us? Is the Earth a very special place? Is our existence here so improbable that we are essentially alone as we stare out at the universe and wonder? Or is the actual scenario somewhere in between?

These are some of the questions to be discussed in this book. Unfortunately, the correct answers for most of them are unknown at this time, and it will be left to the reader to form his or her own opinions on these issues. It is the hope of the author that such opinions will be based on scientific facts and sound logic rather than wild speculation that disregards our current understanding of nature; it is the goal of this book to present a broad spectrum of observational evidence and scientific principles that can be used to formulate coherent opinions on the various aspects of this subject.

The reader will be led down a variety of paths; some of these will arrive at a single conclusion, but most will present the reader with options. Some of these options may be more entertaining, some more palatable, while still others may be more logical. Ultimately, the reader's views are his or her own. This is not to say that all opinions are equally valid, for this cannot be. (Note that we *can* say that two opposing opinions cannot *both* be correct without knowing for sure which one is closer to the truth.) The reader will often have to choose between exciting, popular ideas with little scientific basis and more mundane, commonplace explanations that do not require any mystical beliefs. Throughout the book, the reader will make numerous choices, which should culminate in a consistent expression of belief – one that the reader may find surprising.

In order to properly explore the subject of extraterrestrial life, the reader will need some familiarity with basic astronomy, chemistry, biology, and physics. However, because this book is intended for use in general education classes, no previous experience in any of these sciences is presumed. Topics from these areas will be introduced as they are needed, and developed only to the degree required for comprehension of the problem at hand. This book is not intended to serve as a complete introductory text for any one science; rather, it assembles and presents the fundamental knowledge needed to explore a particular problem that spans several branches of science.

Ideally, this book will discuss – or at least mention – *all* of the major factors in the debate over extraterrestrial life. Realistically, this goal may be difficult to achieve, simply because it is not immediately obvious (or generally agreed) which factors are *most* important to the existence of extraterrestrial civilizations. A crucial factor might have been overlooked; or future observations could turn up new evidence leading to a completely different solution from any of those described in these pages. (If aliens land on the White House lawn next week, this book will need to be revised.) Such is the nature of science.

In any event, it is hoped that this book will be a reasonable guide to those wishing to participate in the discussion on extraterrestrial life. At the very least, it should give the reader a new perspective on alien invasion movies.

Chapter 1
OUR PLACE IN THE UNIVERSE

In which the problem is stated, the scale of the universe is presented, the boundaries of the search are set, the origin of the universe is discussed, and the cast of celestial characters is introduced.

DEFINING THE QUESTION

As indicated by the title, the general subject of this book is "Life in the Universe". However, for reasons that will soon be made clear, this text will attempt to focus on a somewhat more specific topic – "The Abundance of Extraterrestrial Civilizations". By what steps do we get from "Life in the Universe" to "The Abundance of Extraterrestrial Civilizations"?

We could begin by asking simple questions, such as "Is there life in the universe?", to which the answer would be an obvious 'yes': *we* are life and we exist in the universe. Of course we are more interested in whether there is life *other* than ourselves in the universe, life that does *not* come from Earth but rather is **extraterrestrial** (from beyond Earth). With this modified question – "Is there *extraterrestrial* life in the universe?" – the answer becomes less obvious, and yet is still likely to be positive. There are many forms of life on Earth alone, ranging from very simple one-celled creatures to complex beasts such as ourselves, and it would seem likely that in the vastness of the universe there should be some other place where simple life forms have evolved.

As interesting as it would be to know that some type of pond scum exists on a distant world, such news would be unlikely to fascinate the general public for very long. In fact, the means by which we might determine the existence of extraterrestrial pond scum is by no means clear, as neither we, nor the pond scum (we suspect) have mastered interstellar space travel as yet. Far more intriguing would be the discovery of an extraterrestrial species with a degree of intelligence similar to (or perhaps superior to) our own. Our question then becomes "Is there extraterrestrial *intelligent* life in the universe?"

There may be such life out there, but it will be very difficult to discover *them* unless they have become sufficiently technically adept to send and receive interstellar radio messages or travel among the stars to visit us. Such feats will require not only intelligence but also bodies with appendages that permit the beings to carry out engineering projects, environments that supply the necessary raw materials for these activities, and governments that are capable of organizing individuals to fund and execute such ventures. In short, we are looking for a *civilization* at least as intelligent and technical as we are, one that is capable of interstellar radio communication and possibly some limited space travel, bringing us to "Are there extraterrestrial *civilizations* in the universe?"

Now the question is getting very interesting, but it still is wide open. The universe is very large, and either a single extraterrestrial civilization or many would be sufficient to provide an affirmative answer. What we really want to know is not just whether *any* are out there but how many there might be. Knowing the *abundance* of extraterrestrial civilizations would permit us to estimate how close to us the nearest one might be and whether we might hope to make contact with them. Our question has now become "How *abundant* are extraterrestrial civilizations in the universe?" But what is the universe?

A SENSE OF SCALE

As mentioned above, the universe is a very big place. In order to establish just how big it is, let us consider a few terms that are used to define astronomical scales.

The first of these is **solar system**. 'Solar' refers to the Sun, and the solar system is our Sun and the collection of objects that orbit about it – primarily planets (and the moons that orbit them), asteroids, and comets. The Sun is a star, and we are relatively close to it; the stars are suns and are extremely far away. Each star has the potential to have a planetary system of its own, and we have begun to discover such systems around nearby stars.

The second term is **galaxy**. A galaxy contains an enormous number of stars and their planetary systems, huge clouds of interstellar gas and dust, and the remains of stars that have died, all of which are bound together by gravity. Galaxies vary in size and shape; our galaxy – the **Milky Way** Galaxy – includes a few hundred billion stars, making it considerably larger than the solar system. Figure 1.1 shows a sketch of the nearby Andromeda Galaxy, which is similar to the Milky Way. If we could view the Milky Way from the Andromeda Galaxy, our Sun would be essentially invisible, lost in the glare of the much more luminous stars that produce the bulk of the galaxy's visible radiation.

The last term is **universe**, which very simply includes *everything* there is. Looking out from the Milky Way, we see other galaxies, some relatively nearby (such as the Andromeda Galaxy) and many more remote galaxies, extending as far as the telescope can see. For example, the Hubble Ultra Deep Field image was obtained by pointing the Hubble Space Telescope at a tiny portion of the sky apparently devoid of galaxies and exposing for nearly three months; the resulting image revealed a host of previously unknown galaxies, verifying that the currently observed collection of galaxies is merely the tip of the iceberg. The universe is so large that we observe only a very tiny portion of it. We do not know its full extent, but extrapolation from the more closely studied regions indicates that the universe appears to contain *billions* of galaxies.

Clearly, these three terms are not synonymous, but represent a hierarchy of size. We will be careful in this book to apply *universe*, *galaxy*, and *solar system* in their proper places and to *not* use them interchangeably. The reader is advised to adopt a similar approach.

Sizes and distances in astronomy are generally too large to comprehend easily. To put them in a simpler perspective, let us measure the time it would take light to travel each distance, in the same way that we often refer to the length of an automobile excursion in terms of the time required for the trip. Because light travels at a constant speed (300,000 km/s) through the vacuum of space, we can easily

Figure 1.1: M31 – the Andromeda Galaxy.

express distances in terms of light travel times. For example, the circumference of the Earth is about 40,000 km; light requires (40,000 km) ÷ (300,000 km/s) or about 1/7 second to travel this distance. (Of course, light would not travel in a circle around the Earth, as it prefers to go in straight lines instead.)

In a similar fashion we can find that light takes about 1.3 seconds to travel the 384,000 kilometers between the Earth and the Moon, an interval that produced delays of about 2.5 seconds in the conversations between the Apollo astronauts on the moon and the NASA mission control personnel on the Earth. This effect can also be observed on those television newscasts that utilize communications satellites to relay their signals from distant corners of the globe; the roundtrip travel time using a geo-synchronous satellite in a 42,000-kilometer-radius orbit produces a conversational delay of about one half second between each question and answer. Although noticeable, these delays are not usually long enough to cause any serious problems.

The Sun is about 400 times as far away from Earth as the Moon is, and consequently, sunlight requires about 500 seconds (8.3 minutes) to reach Earth. This means that the Sun actually rises about 8 minutes before we observe sunrise, but again, this generally does not cause any great difficulties in our lives.

The distance between Earth and Mars varies as the two planets orbit the Sun, with light travel times between the two planets ranging from about 4 to 20 minutes. This delay places significant restrictions on the control of spacecraft and landers sent to explore Mars; they cannot be operated in real time from the Earth because the round-trip delay of 8 to 40 minutes is just too long. (Note: Control of spacecraft is usually accomplished through radio signals, which also travel at the speed of light.) Spacecraft must be programmed in advance to obtain particular images, and mobile landers must be able to fend for themselves because the Earth-bound backseat driver is too far removed from the action.

The outer planets are even farther away: Jupiter is about 43 light-minutes from the Sun while the dwarf planet Pluto is about 5.5 light-*hours* away. In the distant future, if we should ever establish a base on Pluto, the radio communications will be extremely tedious, with 11-hour-long gaps between questions and answers. At these distances, normal conversations will be quite impossible (and technically correct science fiction movies will become extremely slow-paced).

Beyond the planets, the solar system is represented by a host of small bodies, including comets and other icy worlds, extending to perhaps 1.5 *light-years* from the Sun – a great distance indeed. In these outer reaches of the solar system, the Sun's gravitational control receives an increasing challenge from neighboring stars: **Alpha Centauri**, the *closest* star system to our Sun is 4.3 light-years away, while Sirius, the brightest star in our night sky is about 8.7 light-years distant. The light from these stars requires several years to travel to Earth, meaning that we observe them now as they were several years ago. Our information about stars is never current, but this is not really a problem for astronomers, who are used to looking backwards in time with every observation.

The several thousand stars visible to the naked eye are spread over a wide range of distances extending up to a few thousand light-years from the Sun; for example, it is 520 light-years to Betelgeuse in the constellation Orion, and 1600 light-years to Deneb in Cygnus. In general the other stars in the Milky Way are quite far from the Sun and also from each other, at least in our corner of the Galaxy. Most of the few hundred billion stars that form the Milky Way are spread throughout a flattened pinwheel shape about 80,000 light-years across. Our solar system is not at the center of this pinwheel, but lies about 25,000 light-years from it. These distances are all so great that light cannot traverse them in less than a human lifetime. As the speed of light marks an upper limit on the rate at which material bodies can travel, this would seem to pose some major difficulties for interstellar spacecraft and their crews attempting to explore the Galaxy. This problem will be further examined in Chapter 8.

Outside the Milky Way there are plenty of other galaxies, but only a few of them are relatively nearby. The Large Magellanic Cloud, one of several satellites of the Milky Way, is about 160,000 light-years from us, and the Andromeda Galaxy, the nearest large galaxy (comparable in size to the Milky Way), is about 3 *million* light-years distant. Other galaxies beyond that include the Pinwheel Galaxy (M101) at 27 million light-years, the Sombrero Galaxy (M104) at 37 million light-years, and M87 at 42 million light-years. Of course, many extremely remote galaxies have been found, at distances up to 10 billion light-years or more – close to the estimated age of the universe. At these distances the time delay becomes more important because stars and galaxies change significantly on time scales of a few billion years. What we see now at these great distances is the way the galaxies were very long ago, soon after their formation in some cases. This backwards look in time provides astronomers with a way to probe the distant past, to determine how the universe has changed over time.

Light travel times that range from a fraction of a second to several billion years give some indication of the scale of the universe we inhabit. Another approach is to model the universe using familiar objects; unfortunately, we lack a set of familiar objects that cover the necessary range of sizes and distances. Instead, we can use the same objects to construct a series of 'leapfrog models' that will take us from the Earth to the distant galaxies.

Consider a golf ball, which has a diameter of about 1.7 inches. If the Earth were the size of a golf ball, the Moon would be a small marble about 4 feet away, and the Sun would be a sphere 15 feet across, located about a third of a mile away.

Now if the *Sun* were the size of a golf ball in Omaha, Earth would be a small sand grain 5 yards away. The orbit of Pluto would define a sphere about 400 yards across, and the realm of the solar system's comets would extend to Minneapolis, Davenport, and Wichita. Alpha Centauri would be another golf ball, located in Salt Lake City.

If *Pluto's orbit* were the size of a golf ball in Omaha, Alpha Centauri would be only a fifth of a mile away, the center of the Galaxy would be in Denver, and the disk of the Milky Way Galaxy would extend from the Mississippi River to the Pacific Ocean. On this scale, the Andromeda Galaxy would lie about a quarter of the way to the Moon.

But if the *Milky Way Galaxy* were the size of a golf ball, the Andromeda Galaxy would be another golf ball only about 31 inches away. The Pinwheel Galaxy would be 8 yards away, and M87 would lie about 12 yards distant. The most remote galaxies that we now observe would be about 2 miles away, but the universe extends far beyond that: those galaxies whose light travel times are greater than the age of the universe are beyond our observational grasp.

Our current capabilities do not allow us to travel at – or anywhere near – the speed of light, and thus the extent of our personal exploration of the universe is severely limited. Humans have managed to visit the Moon in person – a distance of about one light-second – with a travel time of a few days. We have also explored parts of our solar system by sending space probes to planets a few light-hours away, with travel times of a few years. At these speeds, expeditions to the nearby stars – a few light-years away – would require travel times of 10,000 to 100,000 years, which seem prohibitive to beings with life spans of about 100 years. Journeys between galaxies a few million light-years apart would appear to be out of the question (at least for us). Unfortunately, our observational reach far exceeds our transportational grasp.

LIMITING THE SEARCH

Because the universe is far too large for us to observe in its entirety, we can hardly hope to determine just how many extraterrestrial civilizations there may be in it, let alone meet them all. At the other extreme, with our growing knowledge of the other worlds in our solar system, it becomes increasingly apparent that none of them provide homes for other intelligent beings. We are almost certainly the *only* civilization within the solar system, as any others this close by should have made their presence known to us or been discovered by our space probes by now. (Humans have established planet Earth as a source of radio emissions that is unique within the solar system and easily detected from the distance of the other planets. A similar civilization within the solar system would not have been overlooked.)

Because the universe is too big for us to examine thoroughly while the solar system is too small to hide any extraterrestrial civilizations, it seems logical to study an object of intermediate size – the Milky Way Galaxy in which we live. With billions of stars in the Milky Way, there is tremendous potential for extraterrestrial civilizations, some of which might be close enough to us that we could actually make contact with them. Whether extraterrestrial civilizations are *common* or *rare* among the stars of the Milky Way is the main subject of this book.

If our Milky Way can be considered to be a representative sample of all the galaxies, then we might be able to make a statement about life elsewhere in the universe. Unfortunately, we do not know the number of civilizations in the Milky Way, nor do we know whether this number is typical, abnormally high, or abnormally low compared to the other galaxies in the universe. For that, we will need to learn something about the various types of galaxies and how they originate. We will start by examining the beginning of the universe.

THE TALE OF COSMOLOGY

When we look out into space at the galaxies around us, we find that they are all in motion. The motion of each individual galaxy can be resolved into two components: one along our line of sight (the radial velocity) and one perpendicular to our line of sight (the tangential velocity). These two components are measured by completely different methods.

Measurement of the tangential component involves observing the rate of change of a galaxy's position on the plane of the sky. But because the galaxies are so far away, this motion is extremely slow and not easily perceived. The radial component is much more readily determined because motion along the line of sight produces a change in the observed frequency of the light we receive. This **Doppler shift** is relatively easy to obtain for objects that are sufficiently bright, and astronomers have used this tool to measure radial velocities of stars and galaxies for many decades.

Radial velocities of stars within the Milky Way show that they approach and recede from the Sun in roughly equal numbers, an appropriate result for a random distribution of velocities. (This velocity distribution is not completely random however, as the stars have some ordered motion about the center of the Galaxy.) One might expect to find a similar arrangement of the velocities of galaxies, but this is not the case; all except our very closest neighbors are moving *away* from the Milky Way. Furthermore, there is a link between a galaxy's radial velocity and its distance from us: the more distant galaxies are moving away more rapidly, with the radial velocity being proportional to distance. This result is known as the **Hubble law**, after its discoverer, Edwin Hubble.

While we can explain the observed *stellar* radial velocity distribution in terms of the orbital motions of stars within the Galaxy, the motions of the galaxies are a bit more puzzling. The rushing away of essentially all of the galaxies might seem to be in defiance of the law of gravity, and their rushing away from *us* seems to imply that *we* are somehow special. Neither of these conclusions is particularly attractive to those who work in the field of **cosmology** – the study of the nature, origin, and evolution of the universe.

Cosmologists prefer a universe that obeys the **cosmological principle**, which says that the large-scale view of the universe should be the same from any galaxy inside it. The distribution of the galaxies we observe from the Milky Way should not be dramatically different from that observed by intelligent beings in the Pinwheel Galaxy or in any of the galaxies in the Hubble Deep Field image; in short, there should be no special vantage point in the universe. But, as noted above, the Hubble law would seem to imply that we *are* in a special galaxy, as the other galaxies are rushing away from us. This apparent contradiction disappears if astronomers in the Pinwheel Galaxy (or any other galaxy) would also find the rest of the galaxies rushing away from *them* in accordance with the Hubble law.

But how could that be? How can observers in *any* galaxy see the other galaxies rushing away from them? The general explanation for this observation is that the universe is *expanding*, causing the distances between the galaxies to increase with time. The galaxies themselves do not expand, nor do the stars, planets, and living beings inside them; only the space that fills the universe is getting bigger. As a very simple example, consider a set of buttons sewn on a strip of elastic; when the elastic is stretched, the strip gets longer, and the buttons get farther apart, but they do not get any larger. The expanding universe accounts nicely for our observation of the Hubble law, and it is consistent with the cosmological principle.

But why is the universe expanding? The currently prevailing explanation is called the **Big Bang Theory**. It says that once upon a time, all the matter in the universe was packed very tightly together in a hot, dense knot called the **primeval fireball**. This fireball then began to expand and cool towards its present state; the start of this expansion is what we call the **Big Bang**.

While we cannot directly observe the actual Big Bang event, we can use our understanding of physics to try to determine the physical conditions and the forms of matter that would have been present in the early moments of the universe, which would have been quite different from our present conditions. The primeval fireball was very hot and quite dense, and the matter was not organized as it is today, into galaxies, stars, planets, and people. Instead, the matter in the universe was all in the form of tiny elementary particles: protons, neutrons, and electrons. These particles combine to make the atoms – to be discussed in Chapter 3 – that comprise our current material world. One type of atom is relatively easy to form: because the nucleus (center) of a hydrogen atom is a single proton, the universe formed hydrogen atoms quite naturally, once it had cooled sufficiently.

Radiation was also present in the early universe in the form of high-energy photons, which interacted continuously with the elementary particles. At first, these interactions prevented the formation of atomic nuclei larger than hydrogen as any protons and neutrons that attempted to combine were immediately blasted apart by the radiation. However, as the universe expanded and cooled, the radiation field became less energetic, permitting the formation of some of the smaller nuclei such as helium, by the process of **nuclear fusion**.

Fusion involves the buildup of larger nuclei from smaller ones and/or elementary particles; it requires protons to bond together in the atomic nucleus even though their positive electrostatic charges normally cause them to repel each other. If they can be brought sufficiently close together, protons can be bound by the **strong nuclear force**, which is more powerful than the electromagnetic force at short range. In the early universe, extremely high temperatures provided protons with enough energy to overcome their electrostatic repulsion and give fusion a chance to proceed.

Given enough time, nuclear fusion might have converted much of the universe's initial hydrogen supply into heavier elements; but the rapidity of the expansion following the Big Bang prevented such an occurrence. The cooling that accompanied the expansion of the universe soon deprived the protons of the energy they needed to overcome the electrostatic repulsive force, halting fusion after only about 30 minutes of activity. Even so, about 25% of the mass of the universe was converted into helium during this interval, with most of the remainder left as hydrogen. Hardly any nuclei heavier than helium were produced, due to a lack of stable nuclei at the next stage of fusion. Formation of the heavier elements necessary for life would have to await the arrival of the stars.

By about a million years after the Big Bang, the expansion and cooling of the universe had weakened the radiation field to the extent that nuclei and electrons could combine to form neutral atoms, a step that also made the universe transparent to visible light. (We cannot see backwards in time beyond this point because the earlier universe was opaque.) Another consequence of the continually weakening radiation field was the emerging dominance of matter over radiation. As the universe expanded, it became progressively easier for matter to begin to form large-scale structures. About a billion years after the Big Bang, the universe began to produce some of the familiar types of objects we observe today: huge clouds of atoms began to form into galaxies, and within them, smaller clouds of atoms condensed into clusters of forming stars. But formation of planets such as Earth could not occur until a sufficient abundance of heavier nuclei had been created.

As noted above, all but the very nearest galaxies are observed to be rushing away from us. But will this continue? What is the future of the universe? The basic Big Bang cosmology offers two different models. In the **closed universe**, the galaxies' outward rush will be slowed by their mutual gravitation until the expansion is finally halted, following which a contraction period will begin and continue until everything in the universe smashes together in an event termed the 'Big Crunch'. In the **open universe**, the galaxies' outward rush will be slowed by their mutual gravitation, but the braking will be insufficient to halt the outward motion, and the universe will expand forever.

One way to attempt to determine which of these models is closer to the truth is to measure the average density of matter in the universe: higher density should produce stronger gravitational forces and lead to a closed universe, while lower density and the resulting weaker gravity should produce an open universe. Over the years, measurements based on the number of galaxies visible in a given volume of space have indicated a relatively low density and, thus, an open universe.

The Big Bang cosmology involves a fairly well defined beginning to the universe, and thus should yield a determinable age. The age of the universe – the time since the Big Bang – can be estimated from the observed expansion rates and measured distances of galaxies. For example, knowing that your car is traveling at 60 miles per hour away from your home, which is now 180 miles behind you, would allow you to conclude that you have been traveling for three hours. In a like manner, we can measure the distance and radial velocity of a given galaxy and use them to calculate the time required to achieve the separation between that galaxy and our Milky Way. Naturally, there are complications: your car needed some time to reach its cruising speed and may have had to slow down for a town along the way; similarly, the rate of expansion of the galaxies has not been constant, due to gravitational braking and other factors. Even so, this procedure should give us a reasonable value for the approximate age of the universe; values have typically ranged from 10 billion years for closed universe models to 15 billion years for open models.

Of course, the situation is not really that straightforward. The Big Bang model has occasionally had to be modified in order to account for new evidence. First came the inflationary universe, which included a brief period of extremely rapid expansion very soon after the Big Bang, followed by a gradually decelerating expansion. A more recent development is the discovery that the universal expansion is actually *accelerating*, rather than decelerating as previously thought, due to a mysterious entity called **dark energy**. The nature of this repulsive force that appears to be driving the galaxies apart is not yet understood.

Our discussion of cosmology brought us to the point at which stars began to form in the first galaxies, perhaps a billion years or so after the Big Bang. Much more needed to happen before life could form on Earth, or anywhere else, but those stories will be related in future chapters. For now, let us note the principal relations between the Big Bang cosmology and our search for extraterrestrial civilizations.

First, our most current estimates place the age of the universe at 13.7 billion years. While this seems to us to be a very long time, it is certainly not an *infinitely* long time. The processes involved in producing galaxies, stars, planets, life, and intelligent beings will each have some minimum time requirements.

If any of these processes are extremely long – on the order of billions of years – then the finite age of the universe may limit the abundance of civilizations.

Second, the raw materials of life must be present before life can form. In Chapter 3 we shall investigate the types of atoms that make up our life; the *origin* of these atoms is then important to the development of our life and possibly other life in the universe. When and where do these atoms form? It was noted above that only the two simplest atoms – hydrogen and helium – are formed in significant quantities by the Big Bang. Other types of atoms, including the majority of those needed for terrestrial life, are formed later on inside the stars. The formation of life had to await the production – and subsequent release – of these atoms, and this may also place limits on the abundance of civilizations.

DANCE OF THE GALAXIES

Galaxies are huge collections of matter scattered throughout the universe. We detect them primarily by the *light* they emit, most of which is produced by the enormous numbers – typically billions – of stars residing in each galaxy. In addition to the stars, there are clouds of gas and dust (from which stars form) and presumably planets, moons, asteroids, comets, etc. comprising planetary systems that accompany at least some of the stars. With all galaxies being initially composed of the same elements – those produced by the Big Bang – the potential for life in other galaxies would seem to be at least as great as it is here in the Milky Way. But is it? Are all galaxies created equal?

As we look around the universe at the galaxies within reach of our telescopes, we find that they are not all alike. Astronomers classify galaxies into different groups according to their shapes. **Elliptical galaxies** have elliptical profiles, and presumably the three-dimensional shape called an ellipsoid: from any direction, its profile would be an ellipse. Some of these galaxies – such as the giant elliptical galaxy M87, shown in Figure 1.2a – have circular profiles and perhaps spherical shapes. Other galaxies exhibit spiral patterns of stars, spread throughout a relatively flat disk shape, with a concentration of stars in a nuclear bulge at the center. These **spiral galaxies** vary considerably in appearance due to the size of the nuclear bulge, the tightness of the wrapping of the spiral arms, and the direction from which they are viewed.

a *b*

Figure 1.2: (a) The giant elliptical galaxy M87; (b) The face-on spiral galaxy M101 – the Pinwheel Galaxy.

Spiral galaxies seen *face-on* may appear as giant pinwheels in the sky, such as M101 – shown in Figure 1.2b; those seen *edge-on* – as the Sombrero Galaxy M104 in Figure 1.3a – hide their spiral structures and masquerade as faint, flattened light bulbs. A third class of galaxies has no special structure or common feature to unite them; these **irregular galaxies** tend to be relatively small and not very luminous, making them difficult to spot at large distances. The Large Magellanic Cloud shown in Figure 1.3b is one example.

 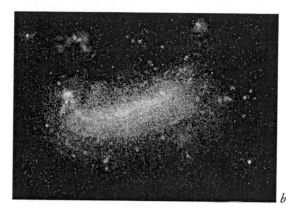

Figure 1.3: (a) The edge-on spiral galaxy M104 – the Sombrero Galaxy;
(b) the Large Magellanic Cloud – a nearby irregular galaxy.

What makes a particular galaxy elliptical, spiral, or irregular? This is not yet fully understood by astronomers, but it may involve several factors, such as the mass of the galaxy, the rate at which it rotates, and its proximity to and interactions with other galaxies. Galaxies seem to be organized into clusters, with individual galaxies bound to a cluster by their mutual gravity. This keeps them relatively close together as they orbit in the cluster as bees around a hive – really big, extremely slowly moving bees, that is. Because of the immense sizes of the galaxies, they occasionally bumble into each other. These interactions may result in mergers of the two participants or mere distortion of their structures.

In either case, the lives of beings in these galaxies may be affected by such events – for better or for worse. Galactic collisions may destroy existing civilizations; alternatively, they may destroy existing dominant-but-not-very-intelligent life forms and thus help to clear the way for evolution of a more intelligent species. We do not know *all* the astronomical events that may have been necessary for our own existence here on Earth, nor do we know what specific astronomical catastrophes may lie in our future, waiting to destroy us. But we should not presume that the universe is generally friendly to and supportive of civilizations such as ours.

We do know that our Milky Way Galaxy is a member of a small cluster of galaxies known (here on Earth) as the **Local Group**. This cluster contains two large spiral galaxies – the Milky Way and the Andromeda Galaxy – and 20 to 30 smaller galaxies of various types. The two large spirals serve as centers of activity with several small galaxies orbiting closely about each. The two best-known satellites of the Milky Way are the Large Magellanic Cloud and the Small Magellanic Cloud; both are irregular galaxies visible as naked-eye objects in the far southern skies. Astronomers have evidence that the Magellanic Clouds interact with our Galaxy, both having passed through the outer part of the Milky Way's disk about 200 million years ago; effects of this passage on civilizations that may have lain in their path have not yet been documented.

GALACTIC REAL ESTATE

As we view it from the Earth, the Milky Way appears as a faint band of light across the night sky. Using telescopes, astronomers have discovered that this light comes from an enormous number of faraway, faint stars, and that the band appearance is due to the flattened shape throughout which the stars are distributed.

The Milky Way is apparently a large spiral galaxy. Most of its stars – estimated at 100 billion to 400 billion in number – reside in the **disk**, an approximately planar region with a diameter of about 80,000 light-years and a thickness of a few thousand light-years. (Figure 1.4 shows an edge view of the disk, along with the other parts of the Galaxy.) At the center of the disk is found the **nuclear bulge**, a spherical region densely populated with stars; in the very center of the Galaxy (the Galactic **core**), a massive **black hole** apparently lurks, devouring whatever stars, gas, and dust dare to approach it. Surrounding the nuclear bulge and the disk is a larger, but less populated spherical volume of stars called the **halo**. If we could view the Milky Way from above the plane of the disk, we would see the **spiral arms** emanating from the nuclear bulge. Our Sun is one of the disk stars, located near a spiral arm about two thirds of the way out from the center to the edge of the disk, a distance of about 25,000 light-years. We live in the suburbs of the Milky Way, far away from the violent stellar neighborhoods found in the more congested regions of the Galactic core.

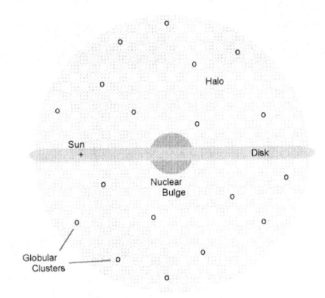

Figure 1.4: Diagram of the Milky Way Galaxy – viewed edge-on.

The stars in the Milky Way are bound together by the Galaxy's gravity, much as the planets in the solar system are bound to the Sun by its gravitational force. And just as the planets orbit the Sun, so too do the stars orbit about the Galaxy's center of mass. Stars in the disk (such as the Sun) follow reasonably circular orbits *within* the disk while stars in the halo plunge right *through* the disk on their more elliptical orbits about the center. (In doing so, they do not collide with the disk stars, due to the extremely large interstellar distances and the relatively small targets the stars present.) The motions of the stars are quite leisurely, at least on a human time scale; the Sun requires about 240 million years to complete one turn about the Galaxy. Stars at different distances from the center will have longer or shorter periods, causing the Galaxy to rotate differentially, rather than as a solid disk. Thus, over very long time-scales, the stars shift with respect to each other, perhaps causing real difficulties for extraterrestrial civilizations attempting to produce accurate navigational maps of the Milky Way.

The Galaxy contains much more than just stars. Some of the stars are organized into *clusters*; astronomers believe that stars *form* in clusters, some of which then drift apart over time while others remain intact. The star clusters that hold together the longest are those with the most stars and hence, the strongest gravity. These **globular clusters** contain around 100,000 stars or so and are distinguished by their roughly spherical appearance and the denser concentration of stars toward their centers. They are distributed throughout the halo and have orbits similar to those of the halo stars. Smaller clusters, with about 100 to 1000 stars, are called **open clusters** or **galactic clusters**; these are found spread along the plane of the Milky Way. In general, they are younger, more recently formed clusters that have not yet had time to disperse. Figure 1.5 shows images of typical star clusters as they appear in a small telescope.

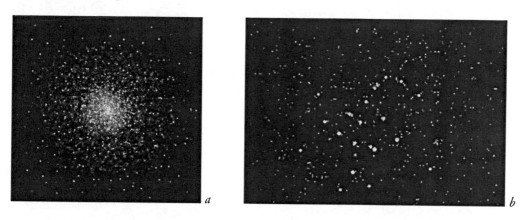

Figure 1.5: (a) A globular star cluster; (b) an open star cluster.

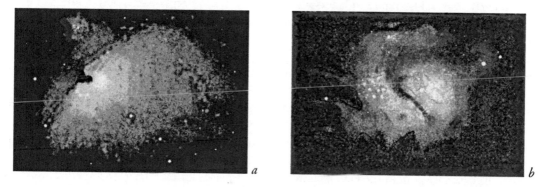

Figure 1.6: (a) M42 – the Orion Nebula;
(b) M8 – the Lagoon Nebula. Both are interstellar gas clouds – regions of star formation.

Between the stars are huge clouds of gas and dust, called **nebulae**. Within some of these nebulae new stars are forming; gravity pulls the individual particles of a cloud together, concentrating them in the center to produce a new star. Figure 1.6 shows examples of two such regions of star formation in the Milky Way.

Within other nebulae, stars are dying. When some stars end their lives, they expel their outer layers back into space, providing the interstellar clouds with more materials to be recycled into new stars. Figure 1.7 shows representatives of the two principal mechanisms for accomplishing this important task – the planetary nebula and the supernova remnant.

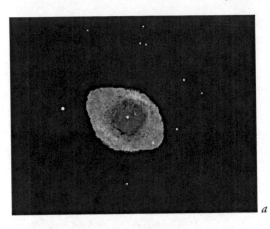 

a *b*

Figure 1.7: (a) M57 – the Ring Nebula (a planetary nebula); (b) M1 – the Crab Nebula (a supernova remnant).

Stars are not all identical; they exist in a wide range of sizes, temperatures, luminosities, and evolutionary stages, as will be discussed in Chapter 6. Some stars still inhabit the clusters in which they were formed, and many stars are actually **binary stars** – systems in which two stars orbit each other, held together by their mutual gravitational forces.

Around at least some – and perhaps most – of the stars are **planets**, generally smaller, cooler bodies, such as the Earth, Mars, and Jupiter. Planets seem to come in a variety of sizes, with a range of different surfaces, as will be seen in Chapter 5. Planets form together with stars, out of the same gas and dust clouds.

Around some planets will be found **moons**, also known as **natural satellites**. These too come in a variety of sizes with many different surfaces, in addition to having different parent planets with which to contend.

Also orbiting about stars will probably be **asteroids** and **comets**; these are leftover materials from the star formation process. Asteroids and comets are both fairly small and quite numerous in our solar system, and we might expect them to be similarly abundant in most other planetary systems – but there are no guarantees. Characteristics of these bodies will be presented in Chapter 5.

Which of these locations in the Galaxy might be good places for extraterrestrial civilizations to live? Our civilization has evolved on a planet, and we normally tend to think of extraterrestrials doing the same, although this may not be the case. If we assume that most species of intelligent extraterrestrial beings will be similar to us in having a *molecular* basis for their structure, we can probably disregard a few classes of objects for very simple reasons: sites that are too hot for most molecules to exist (the surfaces of stars) will clearly be unsuitable; sites where the density of matter is so low that interactions between molecules would be extremely rare (interstellar clouds) will also be less likely to support life. The best environments for life such as ours will probably be those that have suitable temperatures and densities for molecules to exist and interact, on time scales significantly shorter than the lifetimes of the environments. In the next few chapters we will explore the parameters that govern molecules, environments, and their lifetimes.

Main Ideas

- Although the subject of extraterrestrial life is certainly interesting, the focus of this book is on extraterrestrial civilizations with which we might be able to communicate.

- The universe is so immense that humans cannot possibly explore much of it; there may well be extraterrestrial civilizations in it, but we will most likely never find out about them if they exist in galaxies outside ours.

- Our solar system is relatively small, and we have explored enough of it to be quite certain that there are no extraterrestrial civilizations within it.

- Our Galaxy contains a few hundred billion stars, scattered across a visible disk about 80,000 light-years across; it may or may not contain extraterrestrial civilizations close enough to us that we might discover them.

- The Galaxy includes a wide variety of astronomical objects, but only a few of them, such as planets and moons, are likely to serve as sites for intelligent life.

- The universe appears to have begun about 13.7 billion years ago at the time of the Big Bang; thus there has been a limited amount of time available for the development of extra-terrestrial civilizations.

Keywords

Alpha Centauri
Big Bang
black hole
cosmological principle
disk
extraterrestrial
globular clusters
irregular galaxies
moon
nuclear bulge
open universe
solar system
star

Andromeda Galaxy
Big Bang Theory
closed universe
cosmology
Doppler shift
galactic clusters
halo
Local Group
natural satellite
nuclear fusion
planet
spiral arms
strong nuclear force

asteroid
binary star
comet
dark energy
elliptical galaxies
galaxy
Hubble law
Milky Way
nebula
open clusters
primeval fireball
spiral galaxies
universe

LAUNCHPADS

1. Suppose that the Andromeda Galaxy were only 200,000 light-years away. What effects might this have had on the development of life on Earth?

2. If the speed of light were 100 times its present value, would we still place the same limits on our search for extraterrestrial life?

3. Should small galaxies such as the Magellanic Clouds make better, safer homes for extraterrestrial civilizations than large galaxies such as ours?

Chapter 2
THE METHODS OF SCIENCE

In which the principles of the scientific method are explained, options for solving the problem are proposed, the Drake equation is introduced, and applications of probability are discussed.

Chapter 1 presented a brief overview of the universe as we currently envision it. An obvious question to ask at this point is, "How do we know all this?" If the universe is so immense and so much older than our own civilization, how can we hope to gain any significant understanding of its properties and its history? The answer is that by using the methods of science, we can uncover a wealth of information about the universe and its contents. This is not to say that we will learn every possible detail about every feature studied, but the **scientific method** does provide us with a reasonable mechanism for figuring out our environment. Our discussion of this process will begin with a few key terms.

THE SCIENTIFIC METHOD

The first step in trying to understand the universe (or anything else) is to make an **observation**. An observation is simply a statement about an object or event that is witnessed, detected, or measured in some fashion. Such a statement is not necessarily accurate or true because it includes a variety of limitations placed on the observer (by distance, time, etc.), and it depends on the observer's perceptions.

The following are examples of observations:

- The Sun rose this morning.

- Most of the students in this classroom are female.

- Jupiter has four moons.

- A group of lights streaked silently across the sky last night.

All of these observations *could* be true, but none of them are guaranteed. Some of them are true *some* of the time but not *always* (the Sun does not always rise at each location on Earth, each class of students is unique, etc.). Some are partly true (Jupiter has four large, easily visible moons, and a few dozen smaller ones in addition). But each is an observation, made by some person who may or may not have any previous knowledge of or experience with the object or event being observed. Science is based on observations.

We tend to make many observations each day. Those that are routine – I observe that I have five toes on each foot again today – are not given much special attention by our brains. However, those observations that involve something new will usually demand some explanation. A foot with only four toes would set alarm bells ringing in an effort to determine the reason for the change.

The next step in the scientific method is to propose one or more explanations for the observations; each explanation is called a **hypothesis**. Several hypotheses may be devised that can explain the observations equally well (one toe fell off in the night; one toe was gnawed off by the cat; one toe has become invisible; one toe was surgically removed by aliens; etc.). The goal of the scientist is to select which (if any) of these hypotheses is closest to the truth.

Of course, if the original observations are flawed, then the process becomes a wild goose chase. If all five toes are really there (I just miscounted), then there is no need to explain a missing one. Thus, it is especially important to make careful observations and repeat them to be sure they are valid; such well-established observations are called **facts**. Ideally, several different scientists will be able to make the *same* observations, arriving at the *same* results. These facts should provide a firm basis for each hypothesis.

With enough facts at their disposal, scientists are usually able to produce one or more hypotheses, some of which may do a better job of explaining the facts than others. Different hypotheses should also make different predictions about the results of future observations; as more observations are made, certain hypotheses may become stronger while others are abandoned because they are not consistent with the new observations. A hypothesis that becomes well established by this process is called a **theory**.

Occasionally a relatively simple theory will be devised that appears to be generally valid in many situations and passes every test of new observations. Such a *very* well established, simple theory is called a **law**. Laws such as Newton's laws of motion, Kepler's laws of planetary motion, the laws of thermodynamics, the law of gravity, etc., provide the foundation for understanding the world around us.

Observations, hypotheses, facts, theories, and laws are the basis for the scientific method. The goals of science are to discover the basic laws that govern the natural world and to determine which theories will best explain the observational facts.

SCIENCE IN ACTION

As an example, consider a simple observation, easily made by most people: the Sun rises in the east and sets in the west. What hypotheses can be devised to explain this observation?

- Hypothesis #1: The Sun orbits about the Earth, moving from east to west across our sky as it circles around our planet.

- Hypothesis #2: The Earth rotates (spins), with its surface moving from west to east, causing the Sun to *appear* to move in the opposite direction.

- Hypothesis #3: The Sun is carried westward across the sky by invisible angels during the day and hauled underground back to the east by giant turtles every night.

Each of these hypotheses (or some variation of them) has been used to explain the apparent motion of the Sun across our sky. Although it was not immediately obvious to everyone when it was first proposed, the best hypothesis has turned out to be #2.

For another example, raise a book above a table and release it; you should observe the book falling toward the table. Again, several explanations are possible to explain the motion of the book:

- Hypothesis #1: Something below the book – the table, the floor, the Earth – *pulled* it downward.

- Hypothesis #2: Something above the book – the clouds, the Sun, the air – *pushed* it downward.

- Hypothesis #3: The book has an internal propulsion system that propelled it downward.

- Hypothesis #4: The book and the table approach each other and collide due to the curvature of space.

- Hypothesis #5: The book is controlled by extraterrestrial aliens, who caused it to move downward.

In this case, the best answer is not quite so obvious. For many years, the motion of the book was explained by Newton's law of gravity, which says that the Earth and the book attract each other, as described by Hypothesis #1. However, there are many indications that Einstein's theory of general relativity – which says that matter warps space, and objects follow the curvature of space rather than being pulled by some force – is correct. If so, the best explanation would probably be Hypothesis #4.

This brings up two important points. First, a key element in the whole scientific process is that *everything* is open to revision. Better observations may augment or change our collection of facts, and some theories may need to be revised or discarded in favor of those that give a better fit to new observations. Before Copernicus introduced his heliocentric theory of the solar system in 1543, many scholars of the day believed the Earth was the center of the universe. Although not entirely correct, his model has since been shown to be closer to the truth than the older, geocentric theories because it does a better job of explaining the observations.

Occasionally, even a scientific law can be found to be invalid under certain conditions or more limited in its application than had previously been thought, resulting in a revised or new law. This does not mean that *all* scientific knowledge changes every day, but it does mean that any part of it *could* change.

Second, although science attempts to explain many things, it does not *prove* anything. Because everything is subject to revision, nothing can be proven once and for all. It is possible to convince the vast majority of the scientific community of the truth of a particular theory if it is consistent with existing observational evidence and correctly predicts the results of new observations, but absolute proof is impossible to attain. Proof is common in the field of mathematics, but it is not used within the realm of science.

This is all quite acceptable to scientists, but it is often misunderstood by members of the general public who want to know the Absolute Truth. The scientific method is designed to search for the truth, and it may actually attain it in some instances. But science does not know for sure when it has found the truth and cannot *prove* to everyone's satisfaction that it is very close to it. Given the large number of scientific fields, the overwhelming amount of information to be digested in each, and the limited number of scientists available to do the work, it should be understandable that getting close to the truth in as many areas as possible is a fairly reasonable goal.

A Plan of Attack

The question posed in Chapter 1 concerned the abundance of extraterrestrial civilizations in the Galaxy; we would like to determine whether such civilizations exist elsewhere in the Milky Way Galaxy, and if so, just how many there are. How can we find out?

Of course, we could simply *guess* whether or not they exist; the answer is either 'Yes' or 'No', and we have a 50-50 chance of guessing the right answer. But there is no prize for guessing correctly, and we still have no way of knowing whether or not we are right. Guessing the *abundance* of extraterrestrial civilizations is even more difficult because there are so many possible answers from which to choose.

Further, guessing does not contribute any arguments or evidence to the problem and requires no special thought processes for its implementation. In short, it does not get us anywhere.

Perhaps we could use the scientific method – make appropriate observations to gather evidence, form hypotheses to explain the observations, make predictions to test the hypotheses, and ultimately arrive at conclusions about the abundance of extraterrestrial civilizations. With this process we would generate observational facts and logical arguments to use to convince others of the truth of these conclusions. We would still not have any proof (science does not prove things), and our conclusions might not be completely correct (everything is susceptible to revision), but ideally we would be closer to the truth than if we had just guessed.

Our observations might take the form of a search for extraterrestrials. We could conduct a *passive* search, keeping our eyes and ears open and watching for aliens to visit Earth; their physical presence (the alien life forms and the vehicles that brought them here) would provide strong evidence that they exist elsewhere in the Galaxy. Another form of a passive search is to watch for *messages* sent by the aliens, who might use visible light or radio waves to try to communicate with us. An alien message, correctly deciphered and interpreted would also provide evidence of their existence in the Galaxy, although it would probably not be as compelling as their actual presence.

We are already using these passive search techniques. There are thousands of optical telescopes and dozens of radio telescopes around the world that are observing the sky. Although most of them are *not* constantly involved in searching for signals from aliens, there is a reasonable chance that unusual occurrences might not go unnoticed.

Alternatively, we could perform an *active* search. Rather than just watching and waiting, we could travel out into the Galaxy and look for signs of alien civilizations. We could rocket from star to star and planet to planet, looking under every rock and inside every cave in hopes of finding aliens. Such an expedition might turn up *many* civilizations or *none* among the nearby stars likely to be investigated first. Depending on these initial results, a very thorough search of the Galaxy might have to be conducted before we could make a confident statement about the abundance of extraterrestrial civilizations.

Unfortunately, we do not yet have the capabilities to travel around the Galaxy, even to the nearby stars. Even if we did, the Galaxy is so large that such a search would require an extraordinary amount of time, during which new civilizations may have arisen and/or old ones may have died out – possibly even our own. (The problem is somewhat similar to sending a single person out to do a census of the United States by personally counting each citizen.)

It would seem that there is no way we can determine the extent to which the Galaxy is populated with civilizations such as ours, and this may well be the case. After all, we are dealing with an immense region of space containing billions of worlds to be searched; it would be quite understandable if we were to admit that the task is just too overwhelming for us to undertake. However, if we could break the problem down into smaller, more manageable pieces, perhaps we could tackle them one at a time, then recombine the individual results to arrive at a final solution. This is a technique often used for solving everyday problems.

A POTENTIAL SOLUTION

Suppose you are buying pizza for a large group of people; how do you decide how many pizzas to order? Of course you could just guess and hope to come close, but a better number can be obtained by estimating the average amount each person will eat (perhaps one half of a pizza) and multiplying this by the estimated number of people.

Or consider the process of determining the value of a used car. There are a number of factors that will affect the result, such as the make, model, year, mileage, features (air conditioning, power windows, CD player, etc.), and overall condition of the vehicle. By starting with a base price and then modifying it for each factor, one can arrive at a reasonable estimate of the car's value. Every minor detail about the car need not be known as long as the principal factors are included in the estimate.

A similar method exists for estimating the number of technical civilizations currently existing in our Galaxy. The method, devised by Frank Drake in 1961, uses an equation to calculate this number. The equation – generally referred to as the **Drake equation** – expresses the number of technical civilizations as the product of several factors. The factors – each representing a different aspect of the problem – are determined individually and then combined to produce the final result. One of the best properties of the Drake equation is its flexibility; it can be easily modified to suit the needs of the investigator. Several variations, with different sets of factors, can be found in the literature; the version used here is typical.

The goal of the Drake equation is to provide the user with an estimate of the number of technical civilizations currently existing in our Galaxy. This number will be called 'N' in the equation. The value of N should depend on a number of factors, the simplest of which is the number of stars in the Galaxy (N_*). The more stars in the Milky Way, the better the chances of finding life. If each star in the Galaxy supports one civilization, then the Drake equation would be simply $N = N_*$.

However, all stars are not the same; some of them are more suitable for supporting civilizations than others, as will be described in Chapter 6. Obviously the Sun is quite capable of doing the job, and it may be that only those stars that are similar to the Sun will be satisfactory. The number of Sun-like stars in the Galaxy will depend on the total number of stars in the Galaxy (N_*). However, the fraction (f_s) of all stars that are sufficiently Sun-like to support a civilization should be relatively independent of the size of the Galaxy and therefore a quantity that can be estimated. The number of Sun-like stars in the Galaxy will then be the product of N_* and f_s. If each of these Sun-like stars supports one civilization, then the number of civilizations in the Galaxy will be $N = N_* \cdot f_s$ (where the dot (•) signifies multiplication).

As mentioned in Chapter 1, the stars themselves are unlikely sites for life to develop, due to their high temperatures; *planets* that orbit these stars will be much more suitable. The *number* of planets will vary from star to star, and it seems reasonable that the greater the number of planets, the greater the chances of life around a star. The average number of planets per Sun-like star (N_p) will be the next factor in the Drake equation. If N_p is less than one, the number of civilizations will be diminished, and if N_p is greater than one, it will be enhanced. The Drake equation has now become $N = N_* \cdot f_s \cdot N_p$.

Greater numbers of planets should certainly increase the chances of life, but not all planets offer suitable environments for life to survive. There will probably be some limitations on the composition, structure, and temperature of a planet that supports life; perhaps such planets will need to be very similar to Earth (a topic to be explored in Chapter 4). Therefore, the Drake equation requires another factor – f_e, the fraction of planets that are sufficiently Earth-like to support life – resulting in the following form: $N = N_* \cdot f_s \cdot N_p \cdot f_e$.

The above equation would work if every planet that is capable of supporting life actually contains a civilization. However, it is one thing for a planet to be *suitable* for life to exist, but it is another for life to actually *develop* on the planet. The right materials, temperatures, and energy sources on a planet might not guarantee that life will automatically appear there. To account for this possibility, another factor is included: f_l is the fraction of suitable planets on which life actually develops. The equation then becomes $N = N_* \cdot f_s \cdot N_p \cdot f_e \cdot f_l$.

Of course, developing life on a planet does not mean that a technical civilization will appear as well. Humans like to think that such civilizations will arise only from intelligent species. Will a life-bearing planet normally evolve an intelligent species, one that is capable of producing a technical civilization?

Perhaps not, meaning that another factor is needed: f_i is the fraction of life-bearing planets that evolve intelligent beings. The equation now reads $N = N_* \cdot f_s \cdot N_p \cdot f_e \cdot f_l \cdot f_i$.

But evolution of an intelligent species on a planet is still no guarantee that such creatures will develop a technical civilization. The planet they live on will have to supply the necessary raw materials to use in creating and operating a technical civilization; and the intelligent beings will need appropriate body parts in order to assemble the radio telescopes and other supporting equipment needed for interstellar communications. Because these conditions might not be met on every planet with an intelligent species, we must insert still another factor into the equation: f_c is the fraction of intelligent-life-bearing planets on which a technical civilization has developed. The Drake equation now tells us that $N = N_* \cdot f_s \cdot N_p \cdot f_e \cdot f_l \cdot f_i \cdot f_c$.

One might think that this would be good enough – that calculating the number of technical civilizations in the Galaxy is what is desired. However, there is another key factor: time. The Galaxy is very large and very old, but civilizations do not necessarily last forever. Those civilizations that may have arisen in the distant past and then died out will be less interesting to anyone wanting two-way conversations; those civilizations that have not yet achieved the ability to communicate with us will be nearly impossible to find. The Drake equation attempts to estimate the number of technical civilizations *currently existing* in the Galaxy – those that have the capability of interstellar communications at the same time that we do.

The equation given two paragraphs above gives a value for N that is independent of time; it counts all civilizations that *can* arise in the Galaxy (from the present mix of stars and planets), whether they are current, long gone, or still in the future. If technical civilizations last indefinitely once they are formed, this value may be reasonably accurate. But if the lifetime of a technical civilization is quite short, then there is need for another factor in the equation – perhaps f_t, the fraction of time since the development of a planet's initial technical civilization that a technical civilization has existed on the planet. The Drake equation then becomes $N = N_* \cdot f_s \cdot N_p \cdot f_e \cdot f_l \cdot f_i \cdot f_c \cdot f_t$. If technical civilizations have long lifetimes, f_t will be larger, as will N. But if lifetimes are typically short, f_t should be small and N will be correspondingly low.

Estimating f_t can be done by writing this fraction as $f_t = L/t$, where L is the lifetime of an average technical civilization and t is the average time elapsed since the inception of a technical civilization on each planet, which can be estimated at about 3 billion years (see Chapter 7). This is done because the concept of the lifetime of an average technical civilization is somewhat easier to grasp than that of f_t. When this substitution is made, the Drake equation takes on the form that will be employed in this book: $N = N_* \cdot f_s \cdot N_p \cdot f_e \cdot f_l \cdot f_i \cdot f_c \cdot L/t$. (As noted above, there are a number of variations of the Drake equation; most of them – including this one – are somewhat different from Drake's original version. This is not really a problem.)

When viewed as a whole, it can be seen that the equation begins with a large number of stars (N_*) and then whittles this number down by eliminating those stars and planets that will *not* be suitable hosts for technical civilizations. The 'f' factors are fractions, with values between 0 and 1; each factor with a value less than 1 will tend to reduce the estimated value of N. In order to use the Drake equation, one 'simply' estimates values for N_*, f_s, N_p, f_e, f_l, f_i, f_c, and L, inserts them into the equation, and multiplies them all together to determine N. The result obtained depends entirely on the values of the factors used. Because most of these values are not yet known with any great degree of accuracy, the whole problem is wide open for speculation. Different persons using the same Drake equation can come up with completely different – or quite similar – results, depending on what values are inserted for the factors.

With so much uncertainty in the values of the factors, it might seem that the Drake equation is not much help – and there are some who regard it so. However, the Drake equation does provide *some* structure for tackling the problem at hand; it allows the investigator to focus on one particular aspect of the problem at a time and then incorporate the results into a final solution. Its real value is that it makes a very difficult problem more manageable, such that an average reader can estimate values for each factor, plug them into the equation, and determine whether he or she has optimistic or pessimistic views about the existence of extraterrestrial civilizations – a result that may or may not have been obvious before this exercise. The Drake equation will be revisited in Chapter 7, by which time the reader should be ready to estimate the values of several of the factors. A worksheet for filling out the Drake equation is provided in the Appendix.

PROBABILITY AND POSSIBILITY

When using the Drake equation, we will be working with numbers. Some of these will be quite large (such as the number of stars in the Galaxy) while others may be very tiny fractions. It is extremely important that the reader understand the meaning of such numbers and their utilization in calculations. Those needing a refresher course on big numbers and small fractions should see the Appendix at the back of the book.

Many of the small fractions found in this process can be viewed as probabilities. A **probability** is the likelihood of a particular result, such as a visit to the Earth by extraterrestrials in the next year. Probability can be expressed as a number ranging from 0 to 1: if a result has a probability of 0, it has *no* chance of occurring; a result with a probability of 1 should occur every time; and if the probability is 0.5, we can expect the result to occur half of the time.

Some events occur randomly; some do not. Some events are determined by chance; others are not. For example, the result of a dice throw or the winner of a lottery is governed by chance, but the date of the next full moon and your choice of a career are not.

We can attempt to predict the result of an event, whether it is governed by chance or not. It is important to note that the different possible results of a given event are not necessarily equally probable: some results have higher probabilities than others. Furthermore, we may define results in different ways, depending on what our interests are.

A familiar example of a chance event is the flip of a coin. (Additional applications of probability can be found in the Appendix.) Because a normal coin may land either heads up (H) or tails up (T), we can say that the *outcome* of this event will be *either* H or T. And because theses two outcomes are equally likely, the probability of a result of one head landing up (1 h) is 1/2, and the probability of a result of no heads landing up (0 h) is also 1/2.

If additional coins are flipped, the probability of H for *each* coin is still 1/2, but the number of different outcomes increases and the probabilities associated with the various results become more complex. If two coins – a penny and a nickel – are tossed, there are four possible outcomes: both could be heads (HH), both could be tails (TT), the penny could be a head and the nickel could be a tail (HT), or the penny could be a tail and the nickel could be a head (TH). Because each of these outcomes is equally likely, the probability of each is 1/4. However, in terms of the number of heads landing up, there are now *three* different results (2 h, 1 h, and 0 h), and they are *not* equally probable. One outcome (HH) yields two heads, and one outcome (TT) has no heads, but there are *two* different outcomes (HT and TH) that give one head. This means that the probabilities of the three different results are not the same: 2 h should occur 1/4 of the time, 1 h should occur 1/2 of the time, and 0 h should occur 1/4 of the time.

This same analysis can be continued for additional coins. Table 2.1 enumerates the possible outcomes for the tossing of one, two, or three coins and the probabilities for the resulting number of heads.

The total number of outcomes increases as the number of tossed coins increases. The number of different results – the number of heads showing – also increases, but not as rapidly. The probability of each *outcome* is *identical* for a given number of coins, but the probability of each *result* is not necessarily the same. Note that in each case, the total probability for all results is 1.

Table 2.1: Outcomes for Coin Tosses

	1 coin tossed	2 coins tossed	3 coins tossed
Outcomes (results):	H (1 h)	H H (2 h)	H H H (3 h)
	T (0 h)	H T (1 h)	H H T (2 h)
		T H (1 h)	H T H (2 h)
		T T (0 h)	T H H (2 h)
			H T T (1 h)
			T H T (1 h)
			T T H (1 h)
			T T T (0 h)
Total outcomes:	2	4	8
Different results:	2	3	4
	(1 h, 0 h)	(2 h, 1 h, 0 h)	(3 h, 2 h, 1 h, 0 h)
Probabilities of results:	1 h: 1/2 = 0.5	2 h: 1/4 = 0.25	3 h: 1/8 = 0.125
	0 h: 1/2 = 0.5	1 h: 2/4 = 0.50	2 h: 3/8 = 0.375
		0 h: 1/4 = 0.25	1 h: 3/8 = 0.375
			0 h: 1/8 = 0.125

Because the results of coin flips are governed by chance, we would expect the *actual* results to agree with the *predicted* probabilities, as long as enough trials are made. In fact, the *more* trials we make, the *better* should be the agreement between prediction and experiment. If the agreement is not perfect, there is no cause for alarm – random variations are to be *expected* in events of chance.

The probability of throwing *all heads* decreases with the number of coins thrown, as shown in Table 2.2. The probability of throwing *all heads* with 30 coins is 1/1,073,741,824 – about 1 in a billion, a very small number. If 30 coins are flipped a billion times, we could expect to get all heads about once, although any number up to a billion is *possible*. (Only if the probability were 0 would the all-heads result be *impossible*.) Although one in a billion is a small probability, if over a billion trials are made, the result is *likely* to occur. If *several* billion trials are made, the result is likely to occur *several* times.

Note that there is a big difference between the terms *possible* and *probable*. If a result is *possible*, then it *might* happen; if a result is *probable*, then it is *likely* to happen.

Table 2.2: Probability of a Result of All Heads

Number of Coins	Probability of All Heads
1	1/2
2	1/4
3	1/8
4	1/16
5	1/32
n	$1/2^n$

One often hears the phrase "*anything* is possible". This is generally employed to exhort humans to accomplish very difficult tasks, such as graduating from college, beating a superior team, or succeeding in a career, all of which may be possible, depending on the individuals involved. However, the phrase is *not* logically correct and therefore will not be useful in our search for extraterrestrial civilizations. If indeed "*anything* is possible", it should be possible to think of something that is *impossible*. As that would be a *contradiction* – because according to the phrase "*anything* is possible" there should be *no* impossible events – the phrase cannot be true.

There are many things that are, in fact, impossible, although they may not always be obvious. Ideally we would recognize impossibilities as they arise and not waste our time considering them. Realistically it is often difficult to distinguish them from the many low probability events we encounter. In fact, in this book, we will frequently be hard pressed to tell whether an event has a high probability, a low probability, or an *extremely* low probability. And it is this fact that makes the results of the Drake equation so uncertain.

In the next chapter, we will begin to acquire the knowledge needed to make reasonable estimates of factors in the Drake equation. The ultimate goal of this exercise is to determine whether extraterrestrial civilizations in the Galaxy are *probable* or *improbable*; they are almost certainly *possible*.

Main Ideas

- The role of science is to observe and experiment, to develop reasonable hypotheses to explain observations and experimental results, and to determine which hypotheses provide the best explanations for existing facts and make the most accurate predictions of future observations and experiments.

- Science does not *prove* anything, for its findings are always open to revision.

- The Galaxy is so vast that traveling through it to search for extraterrestrial life would be extremely time-consuming and highly impractical – at least for humans.

- The Drake equation is a tool that is designed to provide an estimate of the number of extraterrestrial civilizations currently in our Galaxy; if we can determine reasonable values for the different factors in the Drake equation, we may expect to obtain a reasonable estimate as a result.

Keywords

Drake equation	**fact**	**hypothesis**
law	**observation**	**probability**
scientific method	**theory**	

Launchpads

1. If we had the current capability to travel to the stars, what type of strategy should we adopt to search the Galaxy for extraterrestrials?

2. Perhaps most technical civilizations exist on *moons* rather than planets. How should the Drake equation be modified to address this situation?

3. What type of observational evidence would be most likely to convince the majority of the scientific community that intelligent extraterrestrials exist?

Chapter 3
THE CHEMISTRY OF LIFE

*I*n which atoms and molecules are described, basic chemistry is explained, elemental abundances in terrestrial
*life are reported, reasons for our chemical makeup are explored, DNA and other key molecules of terrestrial
life are described, and implications for the composition and evolution of alien life forms are discussed.*

Life on Earth is based on matter. People, horses, trees, snails, and bacteria are all composed of
matter, as are the Earth, Moon, Sun, planets, and stars. Whether extraterrestrial life will also be
based on matter we do not know, but it is certainly a strong possibility, given our own presence
in the universe. With the broad distribution of matter observed across our Galaxy, it would seem
that plenty of raw materials exist to be assembled into or accessed by extraterrestrial life. But there
are different types of matter, with different properties, and some types may be more suitable than
others for establishing and nourishing life. This chapter will focus on the properties of matter, the
role of matter in terrestrial life, and the similarities we might expect to find between terrestrial and
extraterrestrial life forms.

ATOMS

Any large material object (such as a person, a horse, a tree, a planet, etc.) can be divided into
smaller pieces of matter. This division can be continued until very tiny bits of matter are ultimately
produced. The ancient Greeks believed that there was a limit to this dividing process, reached when
the tiniest bits of matter could no longer be broken down. These smallest possible particles they
called **atoms**.

We still use the term *atom* in a similar sense although atoms are no longer considered to be indivisible
particles. An atom is the smallest particle of a pure substance called an **element**. Atoms of one element are
different from atoms of another element. To see the differences between different atoms, one must examine
their structure.

Atoms can be subdivided, initially into two parts. Each atom contains a hard, dense knot of matter
called the **nucleus**, which is surrounded by a fuzzy region of less dense matter called the **electron cloud**.
Each of these two regions within the atom contains particles. The nucleus contains **protons** and **neutrons**, while the electron cloud contains **electrons**. (Further subdivisions can be made, but these three
particles will be sufficient for our discussion.)

These three elementary particles that compose all atoms each have different properties. The first property of interest is **mass**. The mass of a body is a measure of the amount of matter the body contains. Protons
and neutrons each have a relatively *high* mass while electrons have a relatively *low* mass. The difference is
quite significant: the mass of a proton is about the same as the mass of a neutron, which is about 1836 times
the mass of an electron. This means that essentially all of the mass of an atom resides in the nucleus.

The second property of interest is **electric charge**, and it is different for each of the three elementary particles. The proton is said to have a *positive* charge, and the electron is said to have a *negative* charge. A neutron is electrically neutral and has no charge at all.

The interpretation of positive and negative is as follows. Electric fields exist in many locations; these are *vector* quantities, meaning the field has a *direction* associated with it at each point. A positive charge placed in an electric field will feel a force pushing it in the direction of the field while a negative charge would feel a force pushing it in the opposite direction from the field. A neutral particle would feel *no* electric force.

Charges play an important role in atoms by helping to hold the atom together. Because the nucleus contains protons and neutrons, it will be positively charged while the electron cloud will be negatively charged due to the electrons within. The positively charged nucleus creates an electric field directed *away* from the center of the atom, and this attracts the electrons *toward* the nucleus. (Oppositely charged particles attract each other.) Thus, the atom owes its existence to the properties of its electrically charged constituents.

As mentioned above, atoms are extremely small, having diameters of about one **Ångstrom**. The Ångstrom is a unit of length defined such that a distance of one centimeter – about the width of the nail of your little finger – equals 100 million Ångstroms. Atoms are very tiny indeed. And yet as minute as the atom is, the nucleus is considerably smaller, with a diameter of about 1/100,000 Ångstrom. If the atomic nucleus were the size of a marble (diameter about 1/2 inch), the whole atom would be about a mile and a half across. Most of the *volume* of an atom is occupied by empty space, but most of the mass is confined to a tiny speck in the center. And because a person's body is made of atoms, most of its volume is nothing at all! It would seem that small things could pass right through our bodies without even being detected – and some things actually do!

Human bodies contain an astounding number of these tiny atoms. We can approximate this number by multiplying a person's mass in pounds by 4×10^{25}. Thus, a 150-pound person is comprised of about 6×10^{27} atoms.

ELEMENTS

Different types of atoms are called **elements**; familiar examples include hydrogen, carbon, oxygen, chlorine, iron, gold, lead, and uranium. There are now over 100 elements known, with each element having a unique set of properties, determined by its constituent atoms. But atoms are composed of protons, neutrons, and electrons, particles that are indistinguishable from other particles of the same kind (all protons look alike, etc.). How can so many elements be made from only three different elementary particles? What determines which element will be produced when the three basic particles are combined?

Table 3.1: The Chemical Elements (Emsley 1999)

At. No.	Symbol	Element	At. No.	Symbol	Element	At. No.	Symbol	Element
1	H	Hydrogen	38	Sr	Strontium	75	Re	Rhenium
2	He	Helium	39	Y	Yttrium	76	Os	Osmium
3	Li	Lithium	40	Zr	Zirconium	77	Ir	Iridium
4	Be	Beryllium	41	Nb	Niobium	78	Pt	Platinum
5	B	Boron	42	Mo	Molybdenum	79	Au	Gold
6	C	Carbon	43	Tc	Technetium	80	Hg	Mercury

At. No.	Symbol	Element	At. No.	Symbol	Element	At. No.	Symbol	Element
7	N	Nitrogen	44	Ru	Ruthenium	81	Tl	Thallium
8	O	Oxygen	45	Rh	Rhodium	82	Pb	Lead
9	F	Fluorine	46	Pd	Palladium	83	Bi	Bismuth
10	Ne	Neon	47	Ag	Silver	84	Po	Polonium
11	Na	Sodium	48	Cd	Cadmium	85	At	Astatine
12	Mg	Magnesium	49	In	Indium	86	Rn	Radon
13	Al	Aluminum	50	Sn	Tin	87	Fr	Francium
14	Si	Silicon	51	Sb	Antimony	88	Ra	Radium
15	P	Phosphorus	52	Te	Tellurium	89	Ac	Actinium
16	S	Sulfur	53	I	Iodine	90	Th	Thorium
17	Cl	Chlorine	54	Xe	Xenon	91	Pa	Protactinium
18	Ar	Argon	55	Cs	Cesium	92	U	Uranium
19	K	Potassium	56	Ba	Barium	93	Np	Neptunium
20	Ca	Calcium	57	La	Lanthanum	94	Pu	Plutonium
21	Sc	Scandium	58	Ce	Cerium	95	Am	Americium
22	Ti	Titanium	59	Pr	Praseodymium	96	Cm	Curium
23	V	Vanadium	60	Nd	Neodymium	97	Bk	Berkelium
24	Cr	Chromium	61	Pm	Promethium	98	Cf	Californium
25	Mn	Manganese	62	Sm	Samarium	99	Es	Einsteinium
26	Fe	Iron	63	Eu	Europium	100	Fm	Fermium
27	Co	Cobalt	64	Gd	Gadolinium	101	Md	Mendelevium
28	Ni	Nickel	65	Tb	Terbium	102	No	Nobelium
29	Cu	Copper	66	Dy	Dysprosium	103	Lr	Lawrencium
30	Zn	Zinc	67	Ho	Holmium	104	Rf	Rutherfordium
31	Ga	Gallium	68	Er	Erbium	105	Db	Dubnium
32	Ge	Germanium	69	Tm	Thulium	106	Sg	Seaborgium
33	As	Arsenic	70	Yb	Ytterbium	107	Bh	Bohrium
34	Se	Selenium	71	Lu	Lutetium	108	Hs	Hassium
35	Br	Bromine	72	Hf	Hafnium	109	Mt	Meitnerium
36	Kr	Krypton	73	Ta	Tantalum			
37	Rb	Rubidium	74	W	Tungsten			

The answer lies in the protons, or more properly, the *number* of protons in each atom. It is this number – called the **atomic number** – that identifies the element. Each element has its own unique atomic number, which is equal to the number of protons contained in the nucleus of every atom of that particular element. Therefore an element can be specified by either its name or its atomic number, a point that will come in handy later on. Table 3.1 contains a list of the first 109 elements in order of their atomic numbers.

Another significant atomic property is mass. The **atomic mass number** is equal to the total number of protons and neutrons in an atom (electrons are not counted, due to their very low masses). While the atomic mass number does *not* uniquely define the element, it is still a useful number. Atoms of the same element may contain different numbers of neutrons and therefore have different atomic mass numbers; such atoms are called **isotopes**. Isotopes of certain elements can be important factors in such endeavors as identifying the origins of meteorites, determining ages of materials, or tracing the history of life on Earth.

Each element may exist in nature as several different isotopes. Often there is one predominant isotope, but sometimes there are two or three in significant abundance. Any collection of atoms of such an element will constitute a mixture of different isotopes with different masses. The average mass of these naturally occurring isotopes is called the element's **atomic weight** – a number that is not necessarily an integer.

As an example, consider the element chlorine, which has an atomic number of 17. Every chlorine atom has 17 protons, but they may have different numbers of neutrons. Chlorine's two most abundant isotopes – chlorine-35 and chlorine-37 – have 18 and 20 neutrons, respectively. These isotopes are found in nature in a 3:1 ratio, which yields an atomic weight for chlorine of about 35.5.

The electric charge on the nucleus of an atom is positive and equal in value to the atomic number. The negative charges in the surrounding electron cloud tend to counteract the nuclear charge, reducing the net charge of the atom. In a normal, neutral atom, the number of electrons is equal to the number of protons and the charge on the atom is zero.

Most atoms encountered in everyday life are electrically neutral and thus they neither produce nor respond to electric fields. However, under the right conditions, an atom may lose or gain electrons, becoming an **ion** with an excess charge. For example, atoms in stars can become ionized as the high stellar temperatures pry their electrons free. At more normal temperatures on Earth, atoms in solutions become ionized by exchanging electrons with each other as they dissolve. Atoms in some compounds are held together by ionic bonds produced by the attraction of oppositely charged ions, such as Na^+ and Cl^- in table salt (sodium chloride). The study of the way atoms behave as they bond to each other and dissolve in solutions is called **chemistry**.

Elements can be listed alphabetically or in order of increasing atomic number (as in Table 3.1), but the most common way to display the elements is the **periodic table of the elements**, shown in Figure 3.1. Each box in the periodic table represents a particular element and contains the chemical symbol for the element (H, He, Li, etc.), its atomic number (at the top of the box), and its atomic weight (at the bottom). The convenience of this layout is that it groups the elements according to *both* their atomic numbers and their chemical properties, as will be seen shortly.

Most of the elements on the periodic table can be found in nature, but the heaviest ones – those with atomic numbers greater than 92 – do not occur naturally and must be synthesized in the laboratory. Generally speaking, simpler elements are more abundant in the universe; those elements toward the bottom of the periodic table are rare, mostly because their manufacture is more complicated.

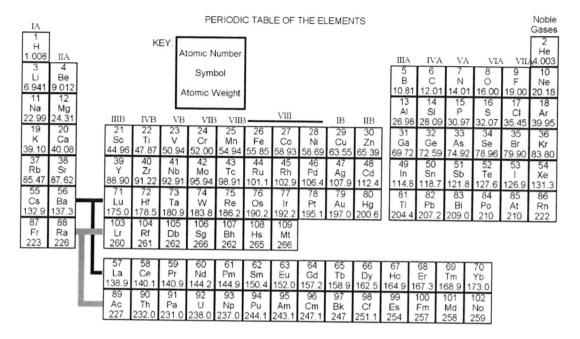

Figure 3.1: Periodic table of the elements (Emsley 1999).

MOLECULES

At relatively low temperatures, atoms combine to form **molecules**. Examples of common molecules include H_2O (water), CH_4 (methane), NH_3 (ammonia), CO_2 (carbon dioxide), and CH_3CH_2OH (ethyl alcohol). The impetus for these combinations is the atom's ambition to obtain eight (or for the smallest atoms, two) electrons in the outer shell of its electron cloud – a more stable configuration. Atoms fulfill this desire by shedding, stealing, or sharing one or more electrons. For example, an atom with two outer electrons makes an ideal molecular partner for one with only six electrons in its outer shell; alternatively, *two* atoms with *one* outer electron each could bond with an atom with six outer electrons (as in the case of water – two of oxygen's eight electrons form its filled inner shell). Each atom has a particular number of electrons and thus forms bonds with other atoms only in certain predictable patterns.

Not all atoms form into molecules. Those atoms that already have a full complement of eight (or for smaller atoms, two) electrons in their outer shells will feel no urge to combine. Such elements – found in the column on the far right side of the periodic table – are known as the noble (or inert) gases because they do not take part in molecular activities. As will be seen later, terrestrial life has a *molecular* basis, and the inert gases (helium, neon, argon, etc.) play no direct role in the structure or function of our bodies.

Many of the molecules that make up terrestrial life are called **organic molecules**; in fact there is a special branch of chemistry (**organic chemistry**) that deals with these molecules. The term 'organic' obviously links these molecules to living organisms, but it has also come to label the particular group of molecules that contain the element *carbon*. More specifically, organic molecules are those containing both carbon and hydrogen, and all terrestrial life is composed of these carbon-based molecules.

ABUNDANCES

It will be important to know which elements make up life forms, planets, stars, the universe, etc. and in what proportions. The **abundance** of an element can be inferred from observations of stellar and planetary spectra and from analysis of sample material from meteorites, the Moon, etc. To understand elemental abundances, imagine dismantling a given object into its constituent elements and then *counting* the atoms of each element. The **number abundance** of a given element is then given in one of several ways.

When discussing cosmic abundances, it is usually easiest to relate the number of each type of atom to the number of atoms of a particular element, such as hydrogen or silicon. For example, Table 3.2 gives the number abundances in our solar system, relative to silicon; these are read as follows: for every 1 million silicon atoms, there are 31.8 billion hydrogen atoms, 2.21 billion helium atoms, etc.

Table 3.2: Solar System Abundances (Basis: Si = 1,000,000) (Lang 1980)

Element	Abundance
hydrogen (H)	31,800,000,000
helium (He)	2,210,000,000
oxygen (O)	21,500,000
carbon (C)	11,800,000
nitrogen (N)	3,740,000
neon (Ne)	3,440,000
magnesium (Mg)	1,061,000
silicon (Si)	1,000,000
iron (Fe)	830,000
sulfur (S)	500,000

Another way to show the number abundance of an element is to calculate the number of atoms of that element divided by the total number of atoms in the object being analyzed, a ratio that may be expressed as a percentage. An abundance of 20% for element X would mean that 20 out of every 100 atoms in the object is an atom of element X.

For example, carbon is one of the principal elements found in terrestrial life, but it is not the most abundant element in our bodies. That distinction goes to hydrogen (H), with an abundance of about 62%. It is followed by oxygen (O) at 24%, carbon (C) at 12%, and nitrogen (N) at 1.2% (Emsley 1999). Thus, just four different elements account for over 99% of all the atoms in terrestrial life, leaving less than 1% for all the rest. Clearly hydrogen, oxygen, carbon, and nitrogen can be considered the elements most essential to life on Earth.

With about 100 different elements from which to choose, why are our bodies comprised primarily of these four? Is there something special about hydrogen, oxygen, carbon, and nitrogen (HOCN) atoms that makes them superior for building life-forming molecules? If so, will they be equally vital to life on other planets? Or are *all* elements – or at least many – just as capable of doing the job? Is our HOCN basis simply the result of chance (maybe these elements happened to form life on Earth *first*) or does their selection have something to do with their abundances in the environment here on Earth? Would life on another planet that is very different from Earth be made of quite different atoms? These questions must be addressed before we can make any definitive statements about extraterrestrial life. And in order to address them, we must learn a bit more chemistry.

CHEMISTRY IN A NUTSHELL

As mentioned above, due to the electronic structure of each element, atoms tend to form certain numbers of bonds with other atoms. Recall that the inert gases in the far right column of the periodic table do not form *any* bonds, being generally satisfied with their electronic structure. The elements in column VIIA (F, Cl, Br, etc.) have one fewer electron than the inert gases and will happily bond with an element that has one extra electron, such as those in column IA (H, Li, Na, etc.), as in NaCl (sodium chloride). Elements in column VIA (O, S, Se, etc.) need two electrons to fill their outer shells and generally form two bonds, as do those in column IIA (Be, Mg, Ca, etc.), which have two extra electrons beyond a filled shell. Similarly, elements in columns headed by boron (B) and nitrogen (N) typically form three bonds each, and those in the column headed by carbon (C) form four bonds.

Even this very simplified analysis shows that elements form different numbers of bonds. Those elements that form no bonds at all (the inert gases) will be essentially no use to molecular-based life forms, while those that form many bonds, such as carbon, hold a much greater potential for creating the *variety* of molecules that organisms are apt to need.

To see how this works, consider hydrogen, oxygen, nitrogen, and carbon – the key elements in terrestrial life – and the molecules that can be made from them.

Hydrogen forms only one bond at a time. It may bond with itself, making diatomic hydrogen (H-H or H_2) – a very small molecule, or it may attach to some other atom with its single bond. But hydrogen is always a dead end. With only one bond, it cannot lead anywhere else or link together to form chains. Within molecules, hydrogen atoms usually serve as caps on the bonds of other atoms.

Oxygen normally forms two bonds, either with one other atom (as in diatomic oxygen: O=O or O_2) or with two (as in water: H-O-H or H_2O). Oxygen is a very reactive element that can bond with many different atoms, but it can only bond with one or two atoms at a time. Furthermore, chains of more than two atoms of oxygen are unstable; in fact the only two stable molecules containing just oxygen and hydrogen atoms are water and hydrogen peroxide (H-O-O-H or H_2O_2).

Nitrogen, with its three bonds does little better than oxygen. The only stable molecules containing just nitrogen and hydrogen atoms are ammonia (NH_3) and hydrazine (N_2H_4); long chains of nitrogen atoms are not found.

Carbon is a different story entirely. With its four bonds, carbon can combine with four hydrogen atoms to make methane (CH_4). Replacing one of those hydrogen atoms with another carbon (and its accompanying hydrogens) yields ethane (CH_3-CH_3 or C_2H_6). A third carbon in the chain produces propane ($CH_3-CH_2-CH_3$ or C_3H_8). The pattern goes on indefinitely: butane (C_4H_{10}), pentane (C_5H_{12}), hexane (C_6H_{14}), etc. The carbon chains thus formed need not be straight – branching may occur where a carbon atom joins three or four others. Carbons may form double bonds with each other, producing different molecules with different properties. They may form a chain that closes on itself – a ring structure, again with different properties. A wide variety of these hydrocarbons exist, so wide in fact that the actual number of these molecules is unknown. Chains of carbon atoms provide the backbone on which an essentially unlimited number of different molecules can be constructed.

THE SILICON CHALLENGE

The above discussion demonstrates why *carbon* is regarded as the basis for terrestrial life, rather than the more abundant hydrogen and oxygen atoms in our bodies. The chemistry of the carbon atom is superior to that of any other element in the top two rows of the periodic table for making the molecules of life because carbon is the only one of these elements to form four bonds. Of course, the elements *below* carbon can also form four bonds; are any of these likely to serve as the basis for life on another world?

The most promising of these would appear to be silicon (Si), in the row just below carbon. Recall that elemental abundances generally *decrease* toward the bottom of the periodic table. Column IVA is no exception: silicon is 1/12 as abundant in the universe as carbon while germanium (Ge) atoms are only 1/100,000 as numerous as carbon. However, as we have seen, there is more to the story than abundance.

Although carbon and silicon both form four bonds, the bonds are not equal in strength. Bonds may be weak or strong depending on how tightly the atom holds its electrons to itself and with which atoms it bonds. Carbon has the ability to form long chains of its atoms that make molecules sturdy enough to be the structural basis for life. Silicon, while forming the same number of bonds as carbon, does not bond as strongly with itself, making the silicon-silicon bond only about half as strong as the carbon-carbon bond. This means that long chains of silicon atoms will not be as robust as carbon chains and will be more likely to come apart under stress – not a desirable feature for molecules forming skin or muscle. Table 3.3 illustrates the superiority of carbon chains by listing the energy required to break the carbon-carbon bond (and several other bonds for comparison); energies given include averages for several similar molecules and values for a few specific molecules.

Table 3.3: Single Bond Enthalpies (in kJ/mole) (Moore 1972)

Averages	Bond Enthalpy	Specific Molecules	Bond Enthalpy
C-C	348	CH_3-CH_3	368
N-N	161	NH_2-NH_2	243
O-O	139	OH-OH	213
Si-Si	177		

Of course, high bond strength is not *always* desirable. Terrestrial life is based on active chemistry. Within the bodies and cells of all living beings on Earth there are chemical reactions occurring. Molecules are formed, broken, or rearranged, and energy is either stored within molecules or released by the reactions. Terrestrial life depends on this constant energy exchange, which in turn requires molecules that can participate. Those molecules with bonds so strong that they can only be broken under extreme conditions will be unlikely to be useful as energy sources, although they may be employed in structural situations.

Many of the energy-producing reactions in terrestrial life involve **double bonds** between carbon and oxygen; for example, one of the simplest carbon-oxygen molecules is carbon dioxide (O=C=O or CO_2), which contains two double bonds. Carbon dioxide is a common gas, the product of our respiration; plants use it in the process of photosynthesis, where it is combined with water to produce sugars. What is the comparable silicon-oxygen compound and how do its bonds compare in strength with these carbon-oxygen double bonds?

As it turns out, silicon-oxygen double bonds do not even exist! Silicon atoms – and atoms of the other elements that lie in the column below silicon on the periodic table – are *larger* than carbon atoms, and this prevents them from forming double bonds with oxygen. Double bonds involve two types of bonds: a **sigma bond** (the usual single bond, formed by electrons in s-lobes between the two atoms) and a **pi bond** (formed by electrons that are found in the p-lobes above and below the atoms). If two atoms joined by a sigma bond are nearly equal in size, the electrons in their adjacent p-lobes may also combine to form a pi bond, as in the carbon-carbon and carbon-oxygen diagrams in Figure 3.2. However, because silicon atoms are significantly *larger* than oxygen atoms, their p-lobes do not overlap with those of oxygen, and pi bonds cannot form.

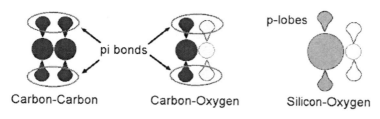

Figure 3.2: The formation of pi bonds from overlapping p-lobes.

There is a compound called silicon dioxide (or silica), but it does not contain any double bonds. In fact, its formula – SiO_2 – does not represent a simple molecule, but rather the empirical formula for a crystalline structure that contains two oxygen atoms for each silicon atom. This is a common substance, existing as quartz, glass, and sand. But while carbon dioxide easily undergoes chemical reactions inside green plants, silicon dioxide is quite inert, making it an excellent material for chemistry glassware. The inability of silicon to form double bonds with oxygen places additional restrictions on the chemistry of the silicon atom that may well make silicon-based life impossible.

Although silicon and carbon both form four bonds and are very close to each other on the periodic table, their atomic properties are dissimilar enough that their chemistry is quite different – certainly different enough that silicon cannot simply substitute for carbon to produce a working system. Add to this the difference in abundance between the two and it is fairly clear why carbon is the backbone element in our bodies. Carbon appears to be uniquely suited to forming the molecules of life here on Earth, and it is difficult to conceive of extraterrestrial life using anything else in its place. It is very tempting to adopt **carbon chauvinism** – the attitude that because *we* are made of carbon, extraterrestrials should also be made of carbon. Of course it may be true, but with the cause and effect switched around. We may be made of carbon *because* it is the system that works best (if not the *only* system), in which case we could logically expect extraterrestrials to have a similar basis (if they exist). Carbon could well be the universal backbone element for chemical life.

THE LIQUID OF LIFE

Life on Earth needs more than just structural molecules and sources of energy. Complex beings such as humans require a transport system to provide each living cell with nutrients and remove waste products effectively. Without miniature conveyor belts, our bodies rely on a fluid (such as a liquid or gas) to move materials from place to place. In addition, cells need a medium in which chemical reactions can be carried out continuously and efficiently, one that holds reacting molecules relatively close together while still allowing them to move around. As gases tend to expand to fill whatever space is available to them, the best choice for these purposes is a liquid – ideally one that dissolves a variety of substances.

A few very simple arguments can be made at this time, to be kept in mind as this investigation proceeds. First, whatever the liquid is, it should be made from relatively abundant atoms. The materials of life should be common, rather than exotic. Second, the molecule that forms the liquid should be small, in order for it to be easily synthesized from its constituent atoms, and also to facilitate the molecule's passage through various membranes as needed. And third, the molecule should be relatively stable, rather than reacting or dissociating at every opportunity.

The liquid of life in our bodies – and the rest of terrestrial life, too – is water. Water easily passes the first three requirements: hydrogen and oxygen are two of the three most abundant elements in our solar system (and in the universe), making the raw materials of water fairly widespread; water is quite small, containing only three atoms; and water is stable enough to form long-lasting oceans covering most of Earth's surface. But is our utilization of water a matter of chance or is it really the best liquid for the job? Does water have properties that make it unique among the many other liquids, much as carbon appears to be unique among the elements? To answer these questions, we must examine the role that water plays in our bodies, consider the properties that make this role possible, and compare these with the properties of other representative liquids.

For this comparison, let us select a variety of liquids; some of them have been suggested as alternatives to water in living systems, while others are included simply to illustrate the range of values for the various properties being examined. Table 3.4 presents a list of these representative molecules, in order of increasing molecular weight. Abbreviations are included for use in subsequent graphs.

Table 3.4: Representative Liquids (Lide 1998)

	formula	abbreviation	molecular weight
methane	CH_4	Met	16.04
ammonia	NH_3	NH3	17.03
water	H_2O	H2O	18.02
formaldehyde	H_2CO	For	30.03
ethane	C_2H_6	Et	30.07
methanol	CH_3OH	MOH	32.04
hydrazine	N_2H_4	Hyd	32.05
hydrogen peroxide	H_2O_2	HP	34.01
propane	C_3H_8	Pro	44.09
ethanol	C_2H_5OH	EOH	46.07
acetone	CH_3COCH_3	Act	58.08
isopropanol	C_3H_7OH	IPA	60.09
benzene	C_6H_6	Bz	78.11
n-octane	C_8H_{18}	Oct	114.23
bromine	Br_2	Br	159.81
mercury	Hg	Hg	200.61

The astute reader may note that some of these substances are not normally liquid at the temperatures and pressures found naturally on the surface of the Earth, and thus they would be of no use to terrestrial organisms. But there may be life forms on other planets that operate at different temperatures using another fluid in their bodies (if another fluid works better), and we should remain open to such possibilities.

Water is a liquid – at least under the right conditions. Under normal atmospheric pressures, water is liquid at temperatures ranging from 0 to 100° Celsius. Figure 3.3 shows how this range compares with the liquid ranges of our other representative substances.

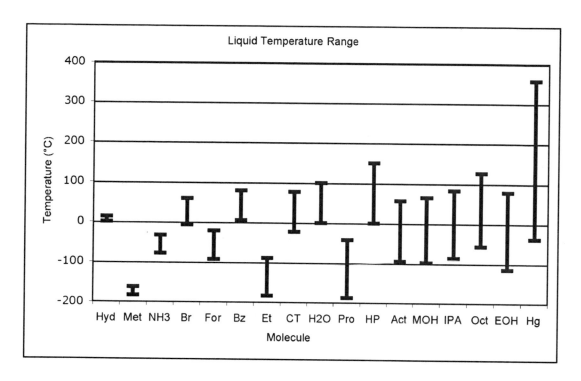

Figure 3.3: Liquid temperature ranges for various molecules.

It should be beneficial for a bodily fluid to have a broad liquid temperature range in order to allow the organism to exist under a wide range of planetary surface conditions. As can be seen by Figure 3.3, water does not have the *greatest* liquid temperature range of this group of molecules, falling instead in the middle of the pack. However, with most of the molecules shown having liquid ranges of 60 degrees or more, it would seem that this property alone does not define a particular fluid as the best for this purpose.

Another factor to be considered is the liquid's performance as a solvent. The degree to which a fluid can dissolve nutrients, waste products, and other materials is crucial to its ability to transport these molecules and ions throughout the organism. Solubility of one substance in another is a complex problem without black-or-white answers, and no simplified treatment will be attempted here. It will be noted that although there is *no* liquid that can dissolve any and all substances, the water in our bodies seems to do a very adequate job of handling the variety of materials needed and produced by living cells. If not a universal solvent, water is certainly very good.

Water in our bodies does much more than just dissolve molecules and distribute them to the cells; it also acts as a thermal regulator, helping to prevent us from getting too hot or too cold. In humans, there are two different properties of water involved: heat capacity and heat of vaporization.

Raising the temperature of any substance generally requires the transfer of heat to it; a thermometer placed in direct sunlight will show an increase in temperature due to the radiant energy it absorbs from the Sun. Similarly, the temperature of water in a pot on the stove will rise as the burners transfer heat to the pot. The *amount* of heat required to increase the temperature of one gram of a substance by one degree Celsius is called the **heat capacity**; different substances have different heat capacities.

Our bodies have evolved to function within a reasonably narrow range of temperatures, and they have defense mechanisms that react automatically to attempt to keep the temperatures of our key organs within this range whenever there is an exchange of heat with the environment. If you step outside without a coat on a cold winter day in Minnesota, your body begins losing heat to the great outdoors; if too much heat is lost, your body temperature will drop too low and your organs may begin to shut down, causing an undesirable end to your life.

How much heat must be lost to lower your temperature out of this range? It depends on the heat capacity of your body's fluid. Our bodies, which are mostly water, require one calorie of heat to change the temperature of one gram by one degree. A fluid with a lower heat capacity would require less heat transfer to change its temperature by one degree, making a body based on such a fluid more susceptible to changes in temperature. A fluid with a high heat capacity will do a better job at stabilizing the temperature of an organism.

The **heat of vaporization** concerns the cooling of a body. Humans gain heat by absorption of sunlight, by absorption of heat rays from objects around them, by contact with warm air, and by chemical reactions in the body that release energy (digestion of food, etc.). We lose heat by similar processes – radiation to the surroundings, contact with cool air, and perspiration. When we become too warm, our bodies secrete sweat (mostly water) onto our skin. If the humidity is not too high, the water then evaporates, changing from liquid into water vapor. This change of phase requires extra energy – the heat of vaporization – to pull the water molecules apart and to create a space for them amongst the air molecules. This energy is supplied by the surface upon which the water resides – our skin. Energy is removed from the skin as the water evaporates, lowering the skin temperature and cooling the body.

In general, the heat of vaporization is the amount of energy required to vaporize one gram of liquid; different substances have different heats of vaporization. A liquid with a *low* heat of vaporization would not work well as a body coolant because more fluid would have to be lost in order to maintain the body's temperature. A liquid with a *high* heat of vaporization is more efficient because cooling can be achieved with minimal expenditure of fluid, giving the body greater potential to preserve itself in hot environments. Thus, a fluid with both a high heat capacity and a high heat of vaporization will be superior as a thermal regulator.

Figure 3.4 illustrates the values of heat capacity and heat of vaporization for our collection of representative liquids. From this data it should be clear that water is unequaled in having *both* a high heat capacity and a high heat of vaporization, making it the obvious best choice of a liquid to function as the thermal regulator within an organism – provided the environment permits it to exist as a liquid.

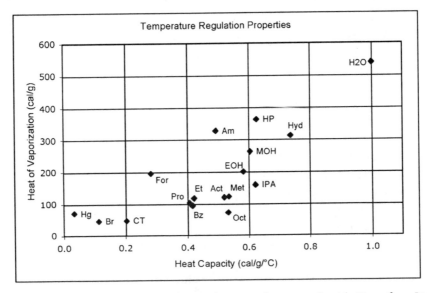

Figure 3.4: Heat capacity and heat of vaporization combinations for various liquids (Data from Lide 1998; Stull, Westrum, and Sinke 1969).

We have already asserted that water is a simple molecule, made from relatively abundant elements. Its simplicity is illustrated by its low molecular weight (water is third lowest on our list, as shown in Table 3.4) and by the small number of atoms per water molecule (3). We will use a slightly more complex scheme to rate molecules according to the abundance of their constituent elements.

The molecules on our list can be formed using only seven different elements: H (1), O (3), C (4), N (5), Cl (19), Br (36), and Hg (60). The number in parentheses indicates each element's number-abundance rank in the universe – hydrogen is first, oxygen is third, etc. An abundance sum can then be formed for each molecule by adding the ranks of each element contained in the molecule; for example, water's sum is 1 + 3 = 4, while carbon tetrachloride's is 4 + 19 = 23. Those molecules with lower abundance sums are made from elements that are more abundant in the universe; our ideal fluid should have a fairly low abundance sum.

Figure 3.5 combines the data for simplicity (using the number of atoms per molecule) and abundance (using the abundance sum) to show which liquids are apt to be good candidates based on these factors. For both factors, a low number is desirable, and water again seems to have the best combination.

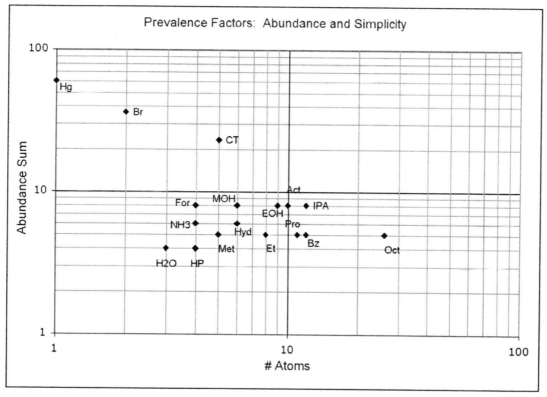

Figure 3.5: Abundance sum vs. number of atoms for various liquids.

With many different factors to compare, it might be difficult to determine which liquid is the best all-around choice for a bodily fluid. In case there is any doubt at this point, we present a simple method of comparison that involves ranking each molecule using each of the various characteristics. The data selected for this ranking include number of atoms per molecule, molecular weight, abundance sum, liquid temperature range, heat capacity, and heat of vaporization. Molecules were ordered within each category and ranked from 1 to 17, with a rank of 1 indicating the best – fewest atoms, lowest molecular weight, lowest abundance sum, largest temperature range, highest heat capacity, and highest heat of vaporization,. The six rankings were then totaled for each molecule, with the lowest total indicating the best molecule.

Results of this exercise are shown in Figure 3.6. Water is easily the best candidate, with its six rankings totaling only 18. While this procedure is fairly crude and ignores data on solubility and stability, it should convey the impression that water is indeed a good choice for a bodily fluid and that few other molecules come close to matching its properties. (Those considering hydrogen peroxide should note that this molecule decomposes into water and oxygen when given a chance.)

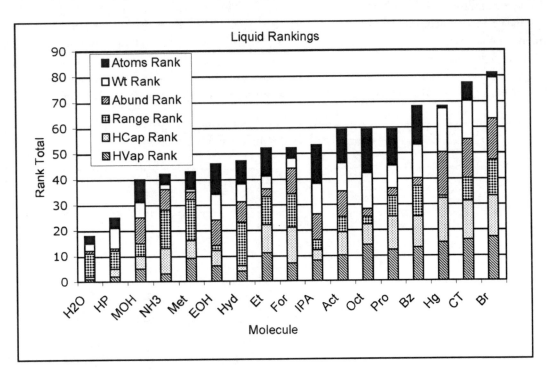

Figure 3.6: Ranking totals for representative liquids (in order, left to right: water, hydrogen peroxide, methanol, ammonia, methane, ethanol, hydrazine, ethane, formaldehyde, isopropanol, acetone, n-octane, propane, benzene, mercury, carbon tetrachloride, and bromine).

In summary, water seems to be an excellent molecule for life forms to utilize as a fluid. It is composed of very abundant elements, it is quite small and easy to form, and it is very stable. In addition, water is liquid over a reasonably wide range of temperatures, is a good solvent for a variety of substances, and has excellent thermal regulation properties with its high heat capacity and high heat of vaporization. All of these features together appear to make water an obvious choice as the basis for a bodily fluid, and it should be no surprise that terrestrial life uses it.

Will extraterrestrial life forms also be based on water? That probably depends on the temperature range available to them; for example, water would not work well on a much colder planet, but ammonia or ethane might. It may be that water is the *only* liquid that works for life anywhere, or perhaps it is simply the best bodily fluid that money can buy. These hypotheses are as yet untested, but a strong case *can* be made for water as the basis of life on planets with temperatures similar to those found on Earth.

ALIEN ABUNDANCES

At this point we return to the subject of elemental abundance: how much of each element is present in a given object? Many of our conclusions will depend on knowing which elements are available in a given environment. Consider first a very simple analysis. Suppose we determine the number abundance of each element (as described above) and then rank the elements in order of decreasing abundance. The top ten elements in the universe (as determined from meteoritic abundances and stellar spectra) and in humans are listed in Table 3.5.

Several points should be noted. First, the principal elements in our bodies (H, O, C, N) are among the most abundant in the universe; we are made of very common materials. Second, the inert gases helium and neon – which are high in abundance in the universe – are *not* utilized in our bodies in any significant capacity. Third, although silicon is relatively high in abundance in the universe, it does not rank among the top ten elements in our bodies, being present only in trace amounts. While high abundance is important, it is not the *main* factor in determining whether a given element will be used; that factor is, of course, the chemistry that the element can do. We are made primarily of the elements hydrogen, oxygen, carbon, and nitrogen because they do the right chemistry and are relatively abundant in the universe.

Table 3.5: Elemental Abundances (Lang 1980, Emsley 1999)

Rank	#1	#2	#3	#4	#5	#6	#7	#8	#9	#10
Universe	H	He	O	C	N	Ne	Mg	Si	Fe	S
Humans	H	O	C	N	P	Ca	S	Na	K	Cl

Will extraterrestrial life be made primarily of these same elements? If not, it will have to be composed of less-abundant elements (which should be more difficult), and it will have to rely on something *other* than organic chemistry for its supply of life-giving molecules (which could well be impossible). It is probably more likely that extraterrestrial life will also be based on hydrogen, oxygen, carbon, and nitrogen for the same reasons that we are – these elements are abundant and their chemistry works.

BUILDING BLOCKS OF LIFE

If they are based on the same elements, will the extraterrestrials use the same *molecules*? Undoubtedly there would be some overlap – perhaps quite a lot – but there are so many organic molecules to choose from that it is likely that there will be variations. Although terrestrial life depends heavily on the simple water molecule, the organic molecules that form the basis of life are quite complicated. Without delving too deeply into the subject of biochemistry, we will examine the molecules that serve as the building blocks of life. Each of these building-block molecules can be polymerized – linked together into long chains to perform specific tasks within the body. Extraterrestrial life may well utilize many of the same building blocks to construct similar – or possibly quite different – polymers.

Amino Acids

The first of these building-block molecules are the **amino acids**. These are organic molecules of a particular type that link together to form polymers (long chains) called **proteins**, found in all living cells. Proteins are the principal material of skin, muscle, nerves, blood, enzymes, and hormones, and thus they play a very important role in terrestrial life.

Amino acids each have two characteristic functional groups attached to the same carbon atom called the α(alpha)-carbon: an amino group (-NH$_2$) and a carboxyl group (-COOH). The other two bonds on the α-carbon join it to a hydrogen atom (-H) and an unspecified radical (-R). Because the radical can be *any* structure, the number of different amino acids appears unlimited. Despite this tremendous variety, terrestrial life manages to function with a relatively small number of amino acids – only twenty. In most of these molecules, the α-carbon bonds to four *different* groups, creating an **asymmetric carbon** (with interesting implications for our search).

Although molecular structures are often drawn out on paper (which is two-dimensional), they really are three-dimensional. In particular, the four bonds of the carbon atom are not planar but instead form the shape of a tetrahedron – a pyramid with a triangular base and three triangular sides. In this picture, the carbon atom resides at the center of the tetrahedron with a bonded group at each vertex.

Figure 3.7 shows two 3D structures of an α-carbon in an amino acid; it is bonded to four different groups – -COOH, -NH$_2$, -H, and -R. In each structure, the carboxyl group is at the top of the pyramid with the radical group at the rear corner of the base (dotted line). In Figure 3.7a, the amino group is on the right front corner while in Figure 3.7b it is on the left. These two molecules have the same functional groups but are mirror images of each other, due to the asymmetric carbon; they are called **enantiomers**.

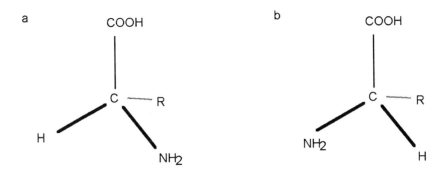

Figure 3.7: Amino acid enantiomers.

By convention, the two structures are labeled according to the position of the amino group: Figure 3.7a, with the amino group on the right is called a D-amino acid, while Figure 3.7b, with the amino group on the left is an L-amino acid. The D- and L- designations are derived from dextrorotary (right) and levorotary (left), terms originally used to describe the direction of rotation of the plane of polarized light by pairs of enantiomers. However, for *these* molecules, D- and L- do *not* indicate rotation direction but simply indicate the amino group position (thus saving the reader from any further explanation of polarized light).

The point of this discussion is that although the enantiomers in each amino acid pair have the same chemical properties and differ only in the arrangement of the functional groups on the α-carbon, they are *not* utilized equally within terrestrial organisms. Terrestrial biology has evolved a system in which *only* the L-amino acids are used to make proteins. Some D-amino acids do exist in life, but they do not participate in the synthesis of proteins.

The reason for this imbalance in roles is not clear. Presumably a biology based on D-amino acids would work just as well, but that is not what we have on Earth. Perhaps there was competition between the two at the inception of life, and the L-amino acid system emerged victorious, completely eliminating the losing system. Or possibly the L-amino acid system was the *only* one to evolve, gaining such an advantage that the other system never had a chance. Will extraterrestrial life use amino acids, and if so, what type will they be?

Molecules from Space

Amino acids have been found in **meteorites**, rocks from space that cross paths with the Earth, streak through our atmosphere, and land on the ground. There are several possible explanations for this observation:

- Hypothesis 1: The amino acids could have been produced here on Earth by terrestrial biology and contaminated the meteorite before it was examined;

- Hypothesis 2: The amino acids could have been produced by non-biological processes occurring somewhere in space;

- Hypothesis 3: The amino acids could have been produced by an extraterrestrial biology somewhere in space.

If the first hypothesis is correct, then the amino acids in the meteorite should all be L-compounds, as found in terrestrial life. If the second hypothesis is right, there should be a *mixture* of L- and D-compounds in roughly equal proportions, as is obtained by synthesizing amino acids in a chemistry lab.

Should the third hypothesis be the best explanation, it is more difficult to predict what form of amino acids would be found. If the extraterrestrial biology is very similar to ours – perhaps even identical – then we might expect only L-amino acids. It is also possible that the alien biology is identical to ours but with the enantiomers reversed, meaning we would find only D-amino acids. Or we could find different amino acids from the 20 that are used by terrestrial life.

Of course the scientific puzzle is worked backwards: we know the result and must determine the cause. If the meteorite contains an equal mixture of L- and D-amino acids, then the odds are good that their origin is non-biological. If only D-amino acids are found, then they were probably produced by an extraterrestrial biology. If we find only L-amino acids, the origin will most likely be biological, but whether it is terrestrial or extraterrestrial will not be easily determined.

To date, the amino acids found in meteorites have included *both* L- and D-compounds in roughly equal proportions, indicating a biological origin is unlikely. Although organic molecules have been found in rocks from space, they do not yet provide direct evidence that extraterrestrial life exists.

Sugars

The next important building-block molecules to consider are the **sugars**, one of the simpler forms of **carbohydrates**, which are so named because the majority of these compounds have an empirical formula of the form $C_n(H_2O)_n$, such as $C_6H_{12}O_6$ (glucose). In terrestrial life forms, sugars are used for energy storage and as structural components. As with the amino acids, the sugars can link together to form polymers. These include starch and cellulose in plants, glycogen and chitin in animals. The sugars also exist as enantiomers, with the

L- and D- nomenclature again used to indicate the relative positions of functional groups (although a D-sugar bears no relation to a D-amino acid). In this case it is the D-sugars that are predominant in nature.

Nucleotides

The last key building-block molecules to be discussed here are the **nucleotides**. These combine to form polymers called nucleic acids, which carry life's genetic code and thus are the basis for the cell's information system. Each nucleotide consists of a sugar, a phosphate group, and a nitrogen base. The sugars derived from the breakdown of nucleic acids tend to be either D-ribose or D-2-deoxyribose; nucleic acid containing the former is called **ribonucleic acid (RNA)** while that containing the latter is **deoxyribonucleic acid (DNA)**.

The nitrogen bases used in the nucleic acids are limited to only five different compounds – adenine, guanine, cytosine, thymine, and uracil, commonly abbreviated A, G, C, T, and U, respectively. Only the first four bases are used in DNA; RNA has the same set but with uracil substituted for thymine.

The polymerization of nucleotides to make nucleic acids results in a chain of alternating sugar and phosphate group units, with the bases sticking out to the side, attached to the sugars. The genetic code is contained in the *order* of these bases; each sequence of three bases is the code for a particular amino acid. The DNA molecule, which carries the genetic code in the nucleus of a cell, is actually a pair of these chains, joined together by links between the bases, much as the rungs of a ladder join the two sides. In DNA, the ladder is also twisted into a spiral, often called a double helix. DNA molecules are enormous, with hundreds of thousands of nucleotides and molecular weights numbering in the millions.

The magic of the DNA molecule occurs in the bonds between the pairs of bases that form each rung on the ladder. The structure of the bases and the geometry of the DNA molecule combine to limit the ways in which the bases can effectively bond. As a result, adenine bonds best with thymine, and cytosine bonds best with guanine. (In RNA, adenine bonds best with uracil, which is very similar in structure to thymine.) This means that the two strands of the DNA molecule form complementary sequences of bases; if one side reads ACGTCTGAA, the other side should read TGCAGACTT.

During replication of the cell, the DNA molecule unzips down the middle to separate the two complementary strands. Each of these strands then reproduces the other side, nucleotide by nucleotide, using components found within the cell. Because of the preferred pairings – A with T, and C with G – the two reproduced DNA molecules will be exact copies of the original, with identical coding carried in each. At least, that is the way it is supposed to work.

In reality, there are occasional errors when the wrong nucleotide is inserted into the sequence. This causes a change in the genetic code that may result in a mutation – a change in the appearance or function of the organism. Mutations are random and will normally be of no particular benefit to the organism. But sometimes the change will result in an improvement, better adapting the organism to its environment or enhancing its ability to grow, feed, and/or elude predators. Life forms that are better suited in some respect will be more likely to survive and reproduce, thus passing along the same mutation to their offspring. In this way, life can evolve to take advantage of changes in the environment or to exploit new ecological niches. Given enough time, a tremendous number of life forms can be developed.

BIOCHEMICAL EXTRAPOLATION

How much time is necessary? Earth formed about 4.6 billion years ago, with the first primitive life forms developing by around 3.5 billion years ago, perhaps in the planet's oceans. For the next 3 billion years, life remained in the sea, experimenting with different forms until some of them emerged onto dry land about 400 million years ago. Since that time essentially all of the species of plants and animals with which we are familiar have evolved, with humans existing for only the last million years or so. Whether life on another world would evolve at the same pace as it has here is unclear; there are a number of random factors, such as asteroid impacts and other astronomical events to be discussed later that can have a dramatic effect on the course of evolution. We do not know whether the time required for the development of intelligent life on Earth was abnormally long, abnormally short, or typical of the time required on other planets.

Not everyone agrees that life on Earth has evolved over billions of years from its primitive beginnings to its current status. Despite the continually increasing fossil record for evolution and our constantly improving understanding of how the numerous fossil pieces fit into the evolutionary puzzle, there are those who prefer to believe that all life was *created* at the beginning of time only a few thousand years ago.

While this belief in creation would seem to present a simpler explanation of the observational facts concerning life on Earth, it still leaves many questions unanswered and creates numerous difficulties in understanding the natural world. Geologists can explain the various formations of rocks found on Earth, but only if they can use *hundreds of millions of years* of plate tectonics and flowing water to sculpt the Earth's surface features. Astronomers can explain the craters on the Moon and the energy sources and life cycles of the Sun and stars, but only if they can have *billions of years* of nuclear fusion and gravitational interactions to process materials and move them around in space. Without these long time-scales, the only alternative is the creationist's magic wand, a solution that is not particularly satisfying to most scientists because it provides no basis for further investigations.

The subject of this book is the existence of extraterrestrial life; what can creationists say about this topic? If creation was responsible for all life on Earth, will it also have produced life on other planets? Or was Earth a special case – the *only* place where life was created? If life was created on Earth, could it still have *evolved* elsewhere? If life was created here and either evolved or was created elsewhere, would all life forms have the same elemental and molecular makeup? Would the intelligent beings created on another planet resemble us in any way? The creationist's answers to these questions are not obvious; it is difficult to make accurate predictions when a magic wand is in control of the outcome. Speculation becomes simply a matter of belief, and those with strong beliefs require little else.

Scientists, on the other hand, make predictions based on observations, hypotheses, facts, and theories. Should the predictions prove to be incorrect, the scientist does not hesitate to modify his or her hypothesis accordingly. Thus, we should be able to reach some conclusions that will have a high probability of being correct based on our knowledge of terrestrial biochemistry:

Will the process of evolution function on other planets? There is no scientific reason why evolution should be unique to Earth. If evolution produced the wide variety of life we find on Earth, it could probably work its magic on another suitable planet, given enough time.

Will extraterrestrial evolution result in beings with the same elemental makeup as we have? It seems reasonable to assume that organic compounds will be involved in alien life, given the unique chemistry of the carbon atom and the abundance of hydrogen, oxygen, carbon, and nitrogen in the universe. The precise abundances will probably vary, but the basic elements will likely be the same.

Will evolution result in beings with the same molecular makeup as we have? Water is very likely to play a key role in life forms on planets where the temperature range permits it to be liquid. Many of the same organic molecules should also be employed in an alien biology, but there is room for plenty of variation.

Will extraterrestrial biology evolve intelligent beings that resemble us? We cannot say whether intelligent extraterrestrials will look anything like humans. In the movies they often do, but that is probably because non-humanoid actors are difficult to find. In reality, the elemental and molecular basis of life does not determine the appearance of the life forms. As a demonstration of this simple fact, consider the vast array of different life forms on Earth, all of which share a common chemistry and molecular basis. Just because we all have similar molecules does not mean that we will resemble each other physically.

Could an extraterrestrial with different chemistry from ours still look like a human? Do intelligent species gravitate toward a certain body plan? Should we expect intelligent aliens to have hearts, brains, eyes, heads, arms, and legs as we do? These are very difficult questions to answer without having met any real extraterrestrials. One could make arguments concerning the necessity or advantage of various body configurations, but that subject will not be covered here.

In summary, we can say that life on Earth is based primarily on chemical elements that are highly abundant throughout the universe. The carbon atom that forms the backbone of all organic molecules seems to be both unique among the elements in its chemical properties and also extremely well suited to forming a host of molecules that are useful to life. Water appears to be a superior bodily fluid, at least at this temperature range. Life on Earth seems to have been based on fairly sound choices; we should not be surprised if extraterrestrial life makes many of the same choices, given a similar environment.

But is a similar environment likely to appear elsewhere in the Galaxy? Is the Earth a unique planet, or are such worlds fairly common in space? What processes had to function in order to produce the Earth that we live on today? These and other questions will be explored in the next two chapters.

MAIN IDEAS

- Most matter with which we are familiar is comprised of atoms; an atom consists of a nucleus of protons and neutrons surrounded by a cloud of electrons.

- The number of protons specifies the type of atom, or element; there are about 100 different elements existing in nature, with each element having a distinctive set of chemical and physical properties.

- Hydrogen and helium make up over 99% of all the atoms in the universe; hydrogen, oxygen, carbon, and nitrogen make up over 99% of all the atoms in terrestrial life.

- Atoms combine chemically to form molecules; the chemical properties of carbon make it particularly well suited to be the basis for terrestrial life, and its relatively high abundance in the universe implies that carbon may also be the basis for extraterrestrial life.

- Water's special chemical and thermal properties make it a superior fluid for living systems, at least in the temperature range over which it remains a liquid.

- Terrestrial life shares a common biochemistry, utilizing the same set of organic molecules – including amino acids, sugars, and nucleotides – to construct complex molecules such as DNA, which provides the basis for our genetic code.

KEYWORDS

abundance	amino acid	Ångstrom
asymmetric carbon	atom	atomic mass number
atomic number	atomic weight	carbohydrates
carbon chauvinism	chemistry	deoxyribonucleic acid (DNA)
double bond	electric charge	electron
electron cloud	element	enantiomers
heat capacity	heat of vaporization	ion
isotopes	mass	meteorite
molecule	neutron	nucleotides
nucleus	number abundance	organic chemistry
organic molecules	periodic table of the elements	pi bond
protein	proton	ribonucleic acid (RNA)
sigma bond	sugar	

LAUNCHPADS

1. If we were to travel to another planet where alien life exists, would we be able to find any food there? Would any of the alien life forms be apt to provide nourishment for our bodies?

2. How would we recognize the presence of extraterrestrial life forms on Earth? What characteristics should these extraterrestrial life forms have that might enable us to distinguish them from terrestrial life forms?

3. Is it likely that alien life could be composed of elements that are completely different from those that make up terrestrial life?

REFERENCES

Emsley, John. 1999. *The elements.* 3rd ed. New York: Oxford University Press.

Lang, Kenneth R. 1980. *Astrophysical formulae.* 2nd ed. New York: Springer-Verlag.

Lide, David R. ed. 1998. *CRC handbook of chemistry and physics.* 79th ed. New York: CRC Press.

Moore, Walter J. 1972. *Physical chemistry.* 4th ed. Englewood Cliffs, NJ: Prentice-Hall.

Stull, Daniel R., Edgar F. Westrum Jr., and Gerard C. Sinke. 1969. *The chemical thermodynamics of organic compounds.* New York: Wiley.

Chapter 4
EVOLUTION OF TERRESTRIAL LIFE

In which the characteristics of the present Earth are reviewed and their relevance to life is noted, methods for estimating ages of rocks and fossils are explained, the principles of evolution are described, and the physical changes in Earth's surface that have accompanied the development of terrestrial life are examined.

The Earth is our home; life formed here, or perhaps formed elsewhere and was transported here. Over billions of years, the Earth and the life it carried evolved together, eventually producing the planet and the life forms we know today. Because our own lives are so short relative to the entire history of the Earth, it is easy for us to think of our surroundings as essentially constant. We may picture the sizes, shapes and positions of continents, oceans, and other geographical features as unchanging over time; similarly, we may regard the numerous species of plants and animals as constants of nature, neither increasing nor decreasing in number. But this viewpoint is simply incorrect. We have ample evidence that dramatic changes have occurred over time – changes in the appearance of Earth's surface, the composition of its atmosphere, the variety of plants that grow in its oceans and on its landmasses, and the types of creatures that run, crawl, slither, fly, and swim across its surface. If we pay close attention, we can observe changes occurring today, as species slide toward extinction, new strains of bacteria are discovered, and continents drift slowly, but measurably, apart. Taken all together, these individual changes – both past and present – are evidence of the evolution of our planet and the life on it. In order to estimate the probability of life on other planets, we must first have some understanding of the properties of the planet on which we live, the history of the evolution of life on Earth, and the manner in which the evolution of life has been intertwined with the development of the Earth itself.

OUR HOME PLANET

Which properties of a planet are necessary for or conducive to the formation of life? So far we have only one example of a planet with life, along with some possibility of finding evidence of life elsewhere in our solar system. Thus, our selection of the crucial characteristics needed for life may be somewhat slanted by our limited experience. Even so, the fact that Earth has life means there must be *something* right about this planet; therefore we will examine the properties of Earth and attempt to determine just what it is that makes our planet special.

Structure

Earth is the prototypical **terrestrial planet**, one of four such planets in the solar system (besides Mercury, Venus, and Mars) and unknown others in the Galaxy. Terrestrial planets are relatively small and are comprised primarily of rock and metal, in contrast to the **Jovian planets**, which are large, gaseous, and made mostly of hydrogen and helium. The interior of the Earth is not uniform in composition, consisting instead of several layers that can be located by studying seismic waves from earthquakes. Earth's central region is called the **core**, divided into a solid inner core and a liquid outer core. This dense core is thought to consist of metallic elements, primarily iron and nickel. Surrounding the core is the **mantle**, composed of rock of lower density than the core materials. The mantle is essentially solid, but it can become plastic and flow under the right combinations of temperature and pressure.

At the surface of the Earth is the **crust** – only a few kilometers thick and composed of rock of lower density than the mantle. This fairly rigid crust includes the continents and the sea floor. Volcanism occurs where the hot mantle material breaks through the crust at active volcanoes or mid-oceanic ridges. The importance of this volcanic action will be discussed later in the chapter.

Above the crust is the Earth's **atmosphere**, a gaseous envelope surrounding the planet. The atmosphere has no distinct upper boundary, gradually decreasing in density with altitude until it becomes imperceptible. Humans exist on the crust at the bottom of this ocean of air; together the crust and the lower atmosphere make up our environment.

Abundances

It is instructive to return to the subject of elemental abundances for a moment. We have seen that terrestrial life is composed primarily of hydrogen, oxygen, carbon, and nitrogen. From where do our bodies obtain these elements? Where in our environment are these elements of life made available?

Table 4.1 gives the number abundances of the principal elements found in a variety of environments. The universe was mentioned in Chapter 3, where it was noted that hydrogen, oxygen, carbon, and nitrogen are among the top five elements in abundance. But we do not have immediate access to all the matter in the universe, and thus it makes more sense to compare our composition with that of the Earth, where we live.

For the Earth as a whole, the top element is oxygen, but this is the *only* one of the four main life elements (H, O, C, N) to make up a significant percentage of the planet. Of course, our environment is not the *entire* Earth; we do not have access to or contact with most of the interior, and it should not be surprising that our bodies do not closely match the abundances found there. Perhaps it would be more appropriate to examine the abundances of the crust, upon which we live.

Crustal abundances are again led by oxygen, with a respectable amount of hydrogen, but carbon and nitrogen are far down the list. This seems a bit odd, that life should have assembled itself out of elements that are not particularly high in abundance on the surface of the planet. Maybe we are still not looking in the right place.

Because life is thought to have originated in the sea, rather than on the crustal rocks, we should probably be looking at the elemental abundances of the oceans. A quick glance at this column shows that hydrogen and oxygen are very high (the sea being made mostly of water), but carbon and nitrogen abundances are again much lower than found in humans.

Table 4.1: Elemental Number Abundances

Universe % (Lang 1980)		Earth % (Baugher 1985)		Crust % (Lide 1998)		Sea % (Lide 1998)		Dry Atmos % (Lide 1998)		Human % (Emsley 1999)	
H	93.4	O	48.9	O	59.9	H	66.2	N	78.4	H	62.2
He	6.49	Fe	18.9	Si	20.9	O	33.1	O	21.1	O	24.1
O	0.0631	Si	14.0	Al	6.34	Cl	0.338	Ar	0.469	C	11.9
C	0.0347	Mg	12.5	H	2.89	Na	0.290	C	0.0159	N	1.15
N	0.0110	S	1.40	Ca	2.15	Mg	0.0328	Ne	0.000913	P	0.226
Ne	0.0101	Al	1.30	Na	2.13	S	0.0174	H	0.000452	Ca	0.224
Mg	0.00312	Na	0.64	Fe	2.10	Ca	0.00635	He	0.000263	S	0.0391
Si	0.00294	Ca	0.46	Mg	1.99	K	0.00631	Kr	0.000057	Na	0.0390
Fe	0.00244	P	0.14	K	1.11	C	0.00144	Xe	0.000004	K	0.0321
S	0.00147	H	0.12	Ti	0.245	Br	0.000520			Cl	0.0240
Ar	0.00034	C	0.099	P	0.0705	B	0.000254			Mg	0.00700
Al	0.00025	K	0.056	F	0.0640	Sr	0.000056			F	0.00123
Ca	0.00021	Mn	0.056	Mn	0.0359	Si	0.000048			Fe	0.00067
Na	0.00018	Cl	0.045	C	0.0346	F	0.000042			Si	0.00032

Of course, another important part of our environment is the atmosphere in which land-dwelling animals and plants are immersed. Could this be the source of our life-giving elements? Here we find plenty of nitrogen and oxygen, along with tiny fractions of carbon and hydrogen – both of which are far below their abundance in our bodies. (Note that the abundances given here are for a standard dry atmosphere; in reality, there is a variable amount of water vapor in the air – up to about 3% by volume – which would slightly increase the amounts of hydrogen and oxygen at the expense of the other elements.)

It is useful to examine the atmospheric composition in terms of molecular abundances, in order to determine the form in which each element is found. In doing so, we find that diatomic nitrogen (N_2) and diatomic oxygen (O_2) comprise the bulk of Earth's atmosphere (at 78% and 21% by volume, respectively), the inert gas argon (Ar) makes up another 0.93%, and carbon dioxide (CO_2) – the chief source of atmospheric carbon – is quite low at only 0.034%. Variable amounts of water vapor and trace amounts of other molecules make up the rest (Cox 2000).

In fact, carbon might appear to be an enigma; with abundances no higher than 0.1% in any of the environments described in Table 4.1, carbon manages to account for about 12% of all the atoms in our bodies. Why does the carbon atom concentrate itself in humans (and other terrestrial life forms)? The answer to this puzzle is of course the magic of carbon chemistry, a chemistry that is unmatched by any other element. Silicon – carbon's closest chemical competitor – despite being second in abundance in Earth's crust is present in our bodies in only trace amounts. It would seem that the existence of carbon as the backbone of terrestrial life – despite its relatively low abundance in our environment – is a very strong indication that extraterrestrial life will also be carbon-based.

Surface Features

The Earth's surface is not uniform, but a combination of oceans and continents. The ratio of dry land to water varies with the average global temperature: when lower temperatures produce ice ages, the sea level drops as more water becomes locked into glaciers and ice caps. Currently the surface distribution is approximately 71% water, 7% ice, and 22% dry land.

Clearly this allocation has worked for life on our planet, but how much leeway do we actually have? Could life have evolved to produce a technical civilization on a planet with only 50% of its surface covered with water? 30%? 10%? 99%? We do not know the acceptable range of values; if it is broad, then many suitable planets may exist, but if it is narrow, we may be alone. A variety of life has existed on Earth for the last few hundred million years, during which time the water/ice/land balance has undoubtedly changed. In the last 100,000 years alone, life forms on Earth – including humans – have endured several ice ages, which dramatically altered the planet's climate. The human race was not extinguished during these difficult times, but the fate of a *technical civilization* confronted with an ice age might have been different. Whether such a civilization could continue to exist by either moving to warmer locations on the planet or somehow halting the advance of the glaciers is unknown.

Other changes in the Earth's surface have occurred – and continue to occur – over time. Recall that the mantle of the Earth is somewhat plastic, allowing it to flow in response to interior pressures. As heat moves outward from the Earth's hot core to the cooler surface, it is transmitted through the mantle by **convection**: hot mantle rock rises while cooler, denser rock sinks. At the surface, the crust is pushed away from these hot spots by the rising mantle material; however, the crust is cooler and more rigid than the underlying mantle such that, rather than flowing smoothly, the crust has broken into a number of plates that ride on top of the gradually shifting mantle. These plates slowly move along the Earth's surface, colliding with, rubbing against, diving under, and/or riding over each other in a process called **plate tectonics**. This process is evidenced by the volcanic activity and earthquakes that accompany the interaction between plates.

At those positions on the planet where the crust is especially thin, molten rock from the mantle can break through the overlying crust onto the sea floor. At the mid-oceanic ridge in the Atlantic Ocean this action is producing new sea floor and shoving the continental plates apart at a rate of a few centimeters per year. This **continental drift** and the plate tectonics theory that explains it received very little support when first proposed by Alfred Wegener in 1915; these ideas did not gain favor until the 1960s, when new evidence on magnetism (see below) forced geologists to revise their thinking.

A few centimeters per year may not sound very significant when discussing the motion of continents separated by thousands of kilometers. But on a time scale of a few hundred million years, those few centimeters per year produce changes on the order of a few thousand kilometers, indicating that our present continental layout has evolved over just such a time scale – in the last ten percent of the Earth's history. Prior to that, the continents existed (for a time) as one giant landmass, which then broke apart into the slowly shifting continents we observe today. The effect of these shifts on the evolution of life will be examined later in this chapter.

The interior of the Earth includes a molten, metallic outer core. Rotation of the Earth causes the charged particles in the fluid core to circulate, which in turn generates the Earth's magnetic field, detectable at the surface with an ordinary compass. The magnetic field is not confined to the interior of the Earth; it extends out into space as well. The magnetic lines of flux (which show the direction a compass needle would align at each point in space) emerge from the surface in the Polar Regions to enclose a large volume of space called the **magnetosphere**, surrounding the Earth out to a distance of several Earth radii. Upon entering the magnetosphere, charged particles from space – mostly electrons and protons from the **solar wind** – feel a magnetic force guiding them toward the Polar Regions. There they enter the atmosphere, colliding with air molecules and causing them to emit light, which we see as an aurora. Some charged particles remain trapped in the magnetosphere where they form the **Van Allen radiation belts**, which occasionally interfere with satellite communications.

The Earth's magnetic field would seem to serve as a protective shield, deflecting many of the charged particles from space away from the equatorial and temperate zones and toward the poles. Has the magnetic field been in any way *responsible* for our presence here? Is a planetary magnetic field *necessary* for the development of intelligent life?

Evidence of changes in the polarity of the magnetic field can be found in the rocks. Magnetic materials in molten rock from volcanic flows tend to align themselves with the local magnetic field, and this alignment is preserved as the molten rock cools and solidifies. Analysis of the (relatively) recently formed sea floor on either side of the mid-oceanic ridge in the Atlantic Ocean shows that the Earth's magnetic field has reversed its polarity many times in the past. Whether these magnetic field reversals were detrimental to, beneficial to, or even noticed by the life forms that experienced them is not known.

Crucial Characteristics

We can enumerate the Earth's characteristics – its interior structure, the composition of its different layers, its surface features, its geologic activity, etc. – but we do not know which of these characteristics have been *necessary* to the development of life here. Most likely the issue is not black or white; certainly some variation in the size of the Earth, the depth of the mantle, the composition of the atmosphere, the strength of the magnetic field, etc. could have been tolerated by an evolving biological system, but we do not know the limits on each factor. This is one of many points where the reader is invited to play the game and make some decisions about which factors are crucial and which ones are superficial.

Of course, it is not all as simple as that, for there is a key factor that has not yet been discussed. The planet described above is the *present* Earth; but life has evolved on Earth over several *billion* years, during which time some of the planet's characteristics have changed dramatically. Before ruling on which factors have been most important, we should investigate what we know of Earth's history. And before we can do that, we should understand how it is possible to establish a chronology of events that occurred before any humans were around to witness them.

AGES OF ROCKS

Our presence on this planet is undoubtedly linked to processes and events that occurred on Earth long before intelligent life appeared here. Then how can we hope to understand the Earth's history without any written record or eyewitness accounts? This problem is not unlike a detective investigating a crime scene, who will seek clues that may provide evidence and then formulate hypotheses to explain what may have happened. Fortunately for us, the Earth has left some clues for us, and they all involve rocks.

Radiometric Dating

The atoms of a chemical element all contain the same number of protons (the atomic number), but they may have different numbers of neutrons. Atoms of the same element that have different numbers of neutrons are called **isotopes**. For most elements, several different isotopes have been studied, although all isotopes are not equally abundant in nature. For example, ^{1}H – hydrogen with one proton, no neutrons, and a mass number of 1 (= #p + #n, indicated by the superscript) – is much more abundant than either ^{2}H (deuterium) with one neutron or ^{3}H (tritium) with two neutrons, with the result that the atomic weight of hydrogen – the weighted average of the atomic masses of the three isotopes – is not much more than one (1.00797). As another example (previously noted), the principal isotopes of chlorine are ^{35}Cl and ^{37}Cl, occurring naturally in an abundance ratio of approximately 3:1, yielding an atomic weight of about 35.5.

One reason for the variation of abundances of isotopes in nature is that different processes produce different isotopes. Some isotopes are synthesized in the interiors of stars, generated by the nuclear fusion reactions that keep the stars shining; others only come into existence through the violence of a supernova – a stellar explosion that liberates enough energy to run a variety of nuclear reactions creating a host of isotopes. A few more are made when cosmic rays bombard atmospheric particles, producing relatively quiet nuclear reactions in our skies. Different nuclear reactions can produce different isotopes, and because the various reactions are not equally probable under the variety of conditions involved, the isotopes they produce are not equally abundant.

Another group of isotopes results from the fact that not all of the nuclei synthesized in the cores of stars, the spectacular fireworks of supernovae, and the cosmic collisions in planetary atmospheres are stable over time. The nucleus of such an unstable isotope may emit a particle, transforming itself into the nucleus of a different isotope or even a different element. In this manner, some isotopes are created while others are destroyed, further affecting the isotopic abundances. These changes are collectively known as **radioactive decay**, and they provide a means for measuring time intervals of different lengths, up to several billion years.

The nature of radioactivity is such that the decay rate of an isotope is proportional to the number of atoms of that isotope that are present. The result is that although the *number* of atoms decaying per unit time interval decreases, the *fraction* of atoms decaying in the same time interval is constant. This allows us to define a **half-life** – the time required for one half of the atoms of a given isotope to decay; half-lives range from tiny fractions of a second (for the very unstable isotopes) to billions of years or more. Some isotopes are stable, lacking a measurable half-life; these are the isotopes normally found in nature.

For example, carbon-12 (or ^{12}C, the nucleus of a carbon atom with six protons, six neutrons, and a mass number of 12) and carbon-13 are both stable isotopes, but carbon-14 is unstable, with a half-life of 5770 years. This means that in a sample of 1 million ^{14}C atoms, 50% will decay over the next 5770 years, leaving 500,000 ^{14}C atoms. In the next 5770 years, another 50% will decay, leaving only 250,000 atoms. At the end of the next several half-lives there will be 125,000 atoms, 62,500 atoms, 31,250 atoms, 15,625 atoms, and so on. After ten half-lives – 57,700 years – fewer than 0.1% of the original ^{14}C atoms will remain. Similarly, aluminum-26, uranium-235, and uranium-238 with half-lives of 740,000 years, 713 million years, and 4.51 billion years, respectively, decay over much longer periods of time.

Collections of radioactive isotopes are thus ticking clocks, with each isotope ticking at a different rate. To measure a given time interval, we use an isotope with a comparable half-life. As an example, consider uranium-235 (the **parent isotope**), which decays through a series of unstable isotopes to lead-207 (the **daughter isotope**), which is stable. For simplicity, assume that at the time of formation of the rock no daughter isotopes are present. As time passes and radioactive decay occurs, the number of parent isotopes declines while the number of daughter isotopes increases. After one half-life (713 million years), there should be equal numbers of uranium-235 and lead-207. After two half-lives (1426 million years), 75 percent of the initial uranium should have decayed, leaving three lead-207 atoms for every uranium-235. We can determine the age of the rock by working the problem backwards: a rock sample containing seven lead-207 isotopes for each uranium-235 must be three half-lives old (2.1 billion years) because the parent isotope has been diminished by a factor of 1/8 (= $1/2^3$). (If daughter isotopes exist in the rock at the time of formation, the problem is more complex, but still solvable.)

Types of Rocks

Radioactive dating thus provides a very convenient method for determining the ages of rock samples; unfortunately, it does not work on all rock types. **Igneous** rocks are formed when magma (molten rock) cools and solidifies. **Sedimentary** rocks are formed by compression and solidification of sediment deposited in layers on the floors of oceans, the particles of sediment originating in the erosion of surface rocks by wind and water. **Metamorphic** rocks are igneous or sedimentary rocks that have been altered (but not completely melted) by pressure and heat associated with tectonic activity. Radioactive isotopes can be employed to determine *absolute* ages of both igneous and metamorphic rocks; the parent/daughter isotope ratios can indicate the time when the igneous rock crystallized or the last time the temperature of a metamorphic rock was high enough to thoroughly mix the isotopes. In each of these cases, the radioactive dating tells how long the rock has been cool enough to retain the parent isotopes and their decay products together in the same location; melting of the rock would scramble the isotopes and thus reset the clock. On the other hand, sedimentary rocks are composed of weathered pieces of many older rocks, each with a different age. Because the processes of weathering and sedimentation do not heat the rock particles and reset the isotopic clocks, radioactive dating is much less useful for sedimentary rocks. These are constructed from countless tiny rock particles, each with its own different age at the time of incorporation into the sediment. Nevertheless, the sedimentary rocks are certainly worthy of study because (1) they form in layers, which can be used to determine *relative* ages of the rock, and (2) they often contain fossils.

Sedimentary rocks form in layers as different types of sediment are carried to the oceans and deposited on the sea floor over millions of years. Following many more millions of years of tectonic activity, the ancient sea beds may be uplifted to become parts of continental plates, where even more millions of years of erosion may reveal their structure. Normally the most recently formed sedimentary rocks are found in the uppermost rock layers while older rocks lie deeper, making determination of relative ages of the different layers fairly straightforward. However, the tectonic forces that move crustal plates around can also deform the rock layers, making horizontal layers become vertical or folding them over to invert the sequence, placing older layers above younger ones. Geologists have been very successful in reading the record of the rocks and interpreting it to determine the sequence of steps that produced each formation and the approximate ages of various layers. This is made easier if the rock formation contains any layers produced by lava flows or magma intrusions – igneous rocks for which *absolute* ages can be obtained. Within this framework, paleontologists (scientists who study fossils) can determine approximate ages for the fossils they find in the various rock layers and can use these to establish the history of life on the planet.

Interpreting the fossil record correctly requires some understanding of the driving force behind evolution and the mechanism by which evolution proceeds. The driving force is linked to environmental changes that have occurred on the Earth's surface over the course of its history, and the evolutionary mechanism is contained in the magic of the DNA molecule. We will now examine the rules by which the game of evolution is played, then move on to follow the environmental changes and their effect on terrestrial life over several billion years.

THE RULES OF BIOLOGICAL EVOLUTION

In the previous chapter we introduced the DNA molecule – the carrier of the genetic code and the blueprint of life. In the nuclei of the cells of your body are DNA molecules that are unique to you, but very similar to the DNA found in every other human. Each human being carries a distinctive set of DNA that defines the characteristics of the individual (red hair, blue eyes, large bones, pale skin, etc.) within the context of the human species: two arms, two legs, no tail, large brain, etc. The cells of other living beings on this planet also contain DNA molecules that are unique to the individual organism but similar to those found in others of the same species. Each cell of an individual of a species contains a certain number of DNA molecules, each of a prescribed length and coded with instructions to manufacture the proteins and other organic molecules needed to construct and operate an individual of that species. The DNA molecules thus characterize the species: human DNA is different from frog DNA and petunia DNA. But although different species have different DNA, these DNA molecules are all based on the *same* set of nucleotides and utilize the *same* genetic code to interpret the information contained in the DNA. All terrestrial life uses the same codebook to store and retrieve DNA information. The information stored varies from species to species, but the code used is always the same.

As noted above, although all individuals of a given species have very similar DNA, the genetic information contained will be slightly different for each individual. And it is these differences that provide the variety we observe within a species. For example, in the human species, genetic differences can account for variations in height, eye color, nose shape, skin color, hair color, etc. These differences are normally passed along to offspring, with the result that children resemble their parents.

Of course, this depends on the method of reproduction. In the case of single cells that reproduce asexually, each DNA molecule makes a copy of itself as the cell divides, with each of the two new daughter cells receiving an identical set of DNA. The new cells are thus clones of the parent cell and should resemble it very closely, assuming their DNA is indeed identical to the original version. However, the DNA copy system is not infallible, and occasional copy errors do occur. Such random errors will change the instructions to the daughter cell and result in characteristics that are different from those of the parent cell. With this system, variations appear randomly in the population, depending on the reliability of the DNA copy system.

In the case of *sexual* reproduction, the procedure is a bit different. The organism produces special cells called **gametes** – eggs or sperm – for use in reproduction, each containing *half* of the normal amount of DNA. Reproduction is accomplished by uniting a sperm from one individual with an egg from another; the resulting cell contains DNA from *both* parents, and it will grow into an individual different from *either* parent. Using this process, variations appear automatically with every reproduction. (Random variations due to copy errors also occur in sexual reproduction, but they are not the principal driving force.) Sexual reproduction is thus a means of enhancing variation within a species population. And variation plays a key role in evolution.

Darwin's Theory

The Earth is populated with a huge variety of plant and animal species, but the extent of this variety was not well understood until comparatively recently. It was not until the early 1800's that humans began to make significant progress in observing and collecting samples of these different species from all around the planet. Most notable of these scientists was Charles Darwin, who studied numerous species of plants and animals during his five-year-long voyage around the world in the 1830's.

In cataloging the species he found on the different continents and islands visited, Darwin paid close attention to both the similarities and the differences he observed. The problem was how to explain the presence of similar species found on disconnected landmasses while also accounting for the observed differences between these similar species. In some cases, particularly in the Galapagos Islands, he found that ecological niches similar to those on the South American continent several hundred kilometers away were occupied by quite different species from those on the mainland – species that were for the most part found only on these islands. For example, of the 26 kinds of land birds found on the islands, 25 were found nowhere else in the world. And of these 25, 13 were species of finches, exhibiting some common structural characteristics but with variations in the sizes and shapes of their beaks. These variations obviously allowed the different finches to exploit different food supplies, a circumstance that permitted a greater number and diversity of finches to exist on the islands than would have been possible if they all competed for the same resources.

Darwin sought explanations for these observations – explanations that would account not only for the unique set of species he found on the Galapagos but also for the observed diversity and distribution of species around the globe. Ultimately, Darwin's explanation was presented in his book *Origin of Species*, published in 1859, in which he described his theory of the origin of species by **natural selection**. The essentials of his theory are as follows:

As living beings reproduce, they generally produce far more offspring than can possibly survive to maturity. (Those species whose offspring normally do *not* live long enough to reproduce will swiftly become extinct.) Although these individual offspring will resemble their parent organisms, there will be some variation, due to the process of sexual reproduction and/or mutations produced by random copy errors in the genetic code. Whatever the cause, the variations will result in organisms that may be better (or worse) suited to living in the current environment. Those individuals that are less well adapted to their situation will have a more difficult time competing for resources, catching prey, and/or avoiding predators and will thus be less likely to survive long enough to reproduce; as a result, the characteristics of such individuals will not be easily retained in the gene pool. On the other hand, individual organisms born with modifications that make them better suited to survival will be more apt to exist long enough to reproduce and pass their improved genes on to their offspring.

In this manner, different species may evolve, fine-tuning themselves to their environments. A subset of a population of one species that is placed in a different environment may evolve to produce a different species. The Galapagos Islands presented such an opportunity when they formed from volcanic activity a few million years ago. Originally lifeless, the islands were colonized by various species of plants, birds, insects, and reptiles that managed to travel there. Over the hundreds of thousands of generations since their arrival, the species involved have become better adapted to the conditions on the islands, to the extent that they have developed new species that are distinct from those found anywhere else. The current inhabitants of the islands should thus resemble their distant cousins on the mainland because they had common ancestors, but they should have evolved into somewhat different creatures because they have been genetically out of touch for so many generations.

Darwin's theory of evolution by natural selection thus provides a way to explain not only the similarities between species but also their differences. The key to the enterprise is time, for genetic changes do not normally occur overnight. For humans living in countries whose histories extend back only a few dozen generations, the effects of hundreds of thousands of generations of evolution may be difficult to fathom; even so, these effects are no less real.

Although the basic mechanism of evolution is reasonably well understood, the *details* of the evolution of terrestrial life are not so clear. This is not surprising; the development of life on this planet has been an extremely complex process, extending over billions of years, and we lack any eyewitnesses or written accounts to paint a complete and accurate picture for us. However, complete, detailed information is normally not required for us to have a reasonable understanding of an event, a process, or an object. We can comprehend the functions of different parts of our bodies without a second-by-second accounting of the status of every cell; we can model the evolution of stars without following the motion of each individual subatomic particle in their interiors; and we hope to be able to trace the history of life on Earth without being familiar with every organism (or even with every species) that has ever lived. More details provide additional observations to be explained by our scientific theories, permitting us to refine and fine tune our ideas about the workings of nature, but we must remember that science and bookkeeping have different goals.

Fossils

If we wish to know how life evolved on this planet, we should study life forms from the past. Unfortunately (for this purpose), the organic material comprising living beings is normally decomposed and recycled fairly rapidly after their demise, leaving little evidence for future study. In some instances bones, teeth, shells, or other hard body parts may be preserved as **fossils**, but the earliest single-celled life forms lacked such parts and thus are not likely to be fossilized. In other cases, the shape of an organism or the tracks it made may be recorded as impressions in mud that eventually hardened into rock. The beds of ancient seas are especially likely to collect fossilized remains as dead marine creatures sink to the bottom and are covered by the sediment that constantly accumulates and eventually forms sedimentary rocks. But unless the local environmental conditions are conducive to the formation of fossils, organisms will live and die without leaving a trace of their existence on Earth.

Even if a fossil *were* formed long ago, its preservation over millions or billions of years until the present would not have been guaranteed. The constant shifting of the Earth's crust due to plate tectonics has erased much of the fossil record; at subduction zones where the less dense continental plates override the seafloor plates, the sedimentary deposits and the fossils they contain plunge back into the mantle, to be melted in the fiery depths. Most of the fossils that do still exist are undoubtedly buried beneath tons of rock on the continents or the seafloors, making them essentially inaccessible to current human investigators. Specialized drills can penetrate deep into the crust to bring up core samples from these hidden layers, but only in regions where rock layers have been lifted, folded, broken, or otherwise exposed by geologic activity can humans easily probe for traces of ancient life.

Evolutionary Details

The fossil record is thus bound to be incomplete, constructed of the fossil remains of a small fraction of the individuals that defined each particular species. Interpretation of this incomplete record has not been straightforward but has depended on the theoretical model of evolution being applied to the data. For many years the standard theory of the evolution of species was that the population of a given species evolved constantly at a very gradual rate, changing over time into a series of different species. New species continually developed from old ones, gradually evolving characteristics that became more and more distinct from those of their ancestors. According to this theory of speciation (formation of new species) – called **phyletic gradualism** – biological structures are continually modified over long periods of time; the fossil record should therefore exhibit examples of the various intermediate organisms from an older species to a newer one, showing bones gradually becoming larger, appendages gradually shrinking, etc. While such a case can be made for some fossil lineages, in many other cases the linkage is not obvious because the fossil record of the intermediate organisms has not been found. This lack of evidence has been used by some to attempt to discredit the whole idea of evolution: if evolution proceeds by gradual changes of species over long periods of time, surely the fossil record should document such a smooth transition with examples of intermediate species. The standard response of proponents of evolution has been to rely on the incompleteness of the fossil record to explain away such gaps; for the various reasons given above, we have not found – nor should we expect to find – complete evolutionary sequences of fossils for every species existing today.

Eldredge and Gould (1972) proposed a different theory of speciation that explains the gaps in the fossil record as real gaps: no fossils of intermediate organisms exist because the evolution of new species is so rapid that the probability of any intermediate organisms being fossilized is extremely small. Under their theory of **punctuated equilibrium**, species do not undergo *gradual* changes over time but rather remain essentially constant (with allowance for some variation within the population). On occasions when a portion of the population of a species becomes geographically isolated from the rest of the population and subject to significantly different environmental conditions, that isolated group may evolve relatively rapidly (on a geologic time scale) into a new species, which will then remain unchanged for long periods of time. The probability of fossilization of an individual of a given species will be proportional to the length of time the species exists; this probability will be relatively high for the established species during its long time of **stasis** between periods of evolution and relatively low for any intermediate forms occurring during its brief evolutionary phase. The general lack of a smooth progression of fossils from one species to another is thus in complete agreement with the predictions of punctuated equilibrium.

If this explanation is correct, it would seem that evolution of new species may occur on demand as needed, rather than continually over time. As long as a species is successful in its environment, it will have no impetus to evolve; the average characteristics of the species will remain essentially constant, with variation around this mean value within a population. When local environmental conditions change, the mechanism of natural selection acts to eliminate those organisms within a population that are less suited to the new conditions and preserve those organisms that are better adapted. The more diverse the population of a species, the greater chance it will include individuals that will be able to adapt to changes in the environment. As an example of environmental changes that can drive evolution, let us return to the topic of plate tectonics that was introduced earlier. What effect has continental drift had on the organisms that ride on these continental rafts?

The Role of Plate Tectonics

The shifting of continents across the Earth's surface has gradually changed their locations and thus their climates as they move in and out of polar or equatorial regions. The breakup of the large land-masses into the present set of continents has also altered the climate by increasing the coastline and thus enhancing the heat transfer rate between the oceans and the dry land. Changes in the sea level produced by plate tectonics have shifted coastlines and forced life forms to adapt to new situations or die out. And the splitting of continents has allowed evolution to proceed in diverse directions, producing different species from the same starting point, due to isolation of groups of plants and animals on each continental raft (as Darwin noticed in the Galapagos).

At first glance it might not be obvious whether plate tectonics and its associated geological and evolutionary changes are *helpful* or *harmful* to the development of technical civilizations. One could argue that the continual changes have been detrimental to the progress of life because changing conditions on Earth have led to the demise of many promising species that just could not adapt fast enough. But it may be that without the driving force of plate tectonics to alter the environment, the evolutionary process would stagnate, advancing at an exceedingly slow pace and taking far too long to produce intelligent life. Will extraterrestrial technical civilizations be able to form on planets that lack plate tectonics? If so, will they form more rapidly or more slowly than here on Earth?

It seems most likely that the net result of plate tectonics will be to *accelerate* the evolution of intelligence and technical civilizations, rather than retard it. The time required for significant changes due to plate tectonics should provide plenty of time for the evolution of new species; for example, at current continental drift rates of two or three centimeters per year, one million years of drift will produce a change of only 20 or 30 kilometers in the gap between continents. While the environmental changes accompanying this 'event' are not apt to be dramatic enough to terminate numerous species on the spot, they *will* provide an impetus for modifications and formation of new species. As continents drift farther apart, differences in their latitudes will translate into differences in the amount of solar radiation they receive. This, together with exposure to different ocean currents will result in different climates for the separating continents, changing the range of ecological niches available on each. Over time, each continent will evolve its own set of plants and animals, producing a greater variety than would have occurred had the landmasses remained in contact with each other. And as the opportunity for development of intelligent species is enhanced by a more diverse population, a process such as plate tectonics that increases diversity may also accelerate the development of intelligence – and any other characteristics that have survival value for a species.

An additional dependence of technical civilizations on plate tectonics (which will be explored later in the book) involves the accessibility of the raw materials that permit intelligent civilizations to become technical. The process of plate tectonics produces mountain ranges, where deposits of metallic ores are often brought to the surface and exposed. Should this *not* happen, due to a lack of plate tectonics, intelligent species might develop and explore an entire planet without ever discovering the metals and other materials from which to build technical devices, such as radios. There may be highly intelligent extraterrestrial beings mired in stone-age civilizations, with no hope of building either radio telescopes or space ships.

Plate tectonics is only one of the mechanisms that has acted to change the surface of the Earth throughout its history. In order to assess the probability of intelligent life on other planets, we must first investigate the development of the Earth's surface and atmosphere over the several billion years since the planet's formation, along with the concomitant evolution of life.

A Brief History of Earth

Earlier in this chapter we presented the basic characteristics of our home planet as we currently observe them. It is tempting to imagine that the Earth has *always* been this way, with our present oceans and continents arranged much as they appear on our globes and a surrounding blanket of air, rich in oxygen and suitable for breathing by humans and other familiar life forms. But the geological record tells us otherwise. Sedimentary rocks found in the interiors of continental landmasses indicate that these areas were once at the bottom of the sea; sedimentary rocks and marine fossils found in mountainous regions tell a similar story. Clearly, plate tectonics has resulted in a bit more than a simple rearrangement of the existing planetary puzzle pieces. Furthermore, we even have evidence of changes in the atmospheric composition over time, based on the oxidation states of metallic ores deposited in different rock layers. The evolution of our planet began at its formation 4.6 billion years ago and continues to this day. What sequence of events has led to our present planetary state?

The First Eon: 4.6–3.6 billion years ago

The Sun, planets, and the rest of the solar system apparently formed from the solar nebula about 4.6 billion years ago – an age inferred from radioactive dating of meteorites. The solar nebula was a mixture of gas and dust, consisting primarily of hydrogen and helium with a very small fraction of heavier elements. The Sun displays this same composition because it developed at the center of the cloud where gravity concentrated most of the gas and dust. Gravitational energy released by the contraction of the cloud heated the forming Sun to stellar temperatures: thousands of degrees on the outside and millions of degrees on the inside. On the other hand, the planets formed from the portion of the solar nebula that did *not* fall into the Sun but rather remained in orbit about it. These gases cooled and condensed to liquid or solid forms, which then began to stick together. Close to the Sun the temperature was relatively high, and only compounds that could condense at such temperatures – primarily metallic and rocky materials – managed to do so. As a result, Earth and the other inner planets have rocky, metallic compositions.

Condensation of these materials produced grains and small particles, which collided and stuck together to form larger and larger particles, then meteoroids, asteroids, **planetesimals**, and ultimately, planets. Throughout this process of condensation and accretion, the number of individual particles orbiting the new Sun gradually diminished as the larger bodies swept up and amalgamated the smaller ones, eventually producing only a few planets throughout the solar system. (The inner planets, being made primarily of the heavy elements that were low in abundance in the solar nebula, remained relatively small compared to the Jovian planets, which were able to acquire and retain hydrogen and helium.) But accretion did not cease once the planets had accumulated most of their mass; asteroids of reasonable size continued to smash into the Earth for hundreds of millions of years after the initial formation. Although both the bombardment rate and the average size of the impacting bodies have declined considerably by now, we continue to be pelted by tiny meteoroids (and very rarely by much larger rocks) as the Earth goes about its orbital business.

Energy from the impacting bodies and the heat released by radioactive decay of unstable elements inside the Earth were sufficient to melt the planet and promote differentiation into an iron core, a rocky mantle, and a slag crust within the first hundred million years or so. Radiative cooling of the exterior would have formed a solid crust over the molten interior, but continued impacts probably prevented a permanent solidification for quite some time.

Impacts played a major role in the Earth's development during this period. In the first hundred million years, a particularly dramatic impact with a planetesimal the size of Mars splashed out material that later formed into our Moon. (The lower density of the Moon compared to Earth is consistent with this theory, which says the lunar material originated from the Earth's mantle *after* differentiation had occurred.) Impacts with other rocky planetesimals added mass to the Earth and formed craters, but all traces of these events have been erased by subsequent geologic activity. Comets that collided with Earth brought volatile materials such as water, to contribute to the atmosphere and eventually the oceans; they may also have delivered organic molecules to the Earth's surface. These impacts and the energy they imparted to the Earth kept melting and re-melting the surface during the period of heavy bombardment, which lasted up to about 3.8 billion years ago, with the result that the oldest Earth rocks known date only to about 4 billion years ago. Thus the rock record from this earliest interval is essentially nonexistent.

The origin of Earth's atmosphere is also difficult to determine with certainty. Some researchers have presumed that the composition of the early atmosphere would have resembled that of the solar nebula, which was abundant in hydrogen, helium, and simple hydrogen-rich compounds of carbon, nitrogen, and oxygen, such as methane (CH_4), ammonia (NH_3), and water vapor (H_2O). However, it is unlikely that such a **reducing atmosphere** was ever present on the Earth because the planet's small mass and resulting low gravity are insufficient to allow hydrogen and helium atoms to be acquired directly from the solar nebula or to prevent them from escaping into space. Instead, the bulk of the original atmosphere may have come from volcanic outgassing, which would have released carbon dioxide (CO_2), diatomic nitrogen (N_2), and water vapor. Impacts by comets comprised of water ice, methane ice, ammonia ice, etc. would have added more volatiles to the atmosphere, including some that might otherwise have been difficult to obtain from the solar nebula. Whether the cometary bombardment provided the forming Earth with the raw materials from which to manufacture life is not clear, but it is certainly a possibility.

As the rate of impacts gradually declined, the surface of the Earth continued to cool. As the atmospheric temperature dropped, water vapor began to condense and fall as rain, and when the surface rocks had cooled sufficiently, the oceans began to form. This cooling process was probably interrupted many times by large impacts that heated and vaporized significant portions of the forming ocean, but finally, by about 3.9 billion years ago, the Earth was covered by an ocean of liquid water.

The topography of the recently molten planet should have been fairly uniform during this time, without the range of variation in altitude that marks the Earth's present surface. Convection within the mantle would have produced hot spots just below the crust, and the resulting volcanic activity would have deposited lava onto the surface in these areas – the beginnings of continental landmasses. However, as the mantle convection was probably not well established at this time, the hot spots would have changed position and the continents could not have grown very large. At best, only scattered, tiny landmasses – protocontinents – could have formed during this period. The early Earth would have been quite unrecognizable to us.

As the oceans formed, the Earth continued to cool by radiating energy into space. The surface temperature of a planet depends on its energy balance – the relative amounts of energy being absorbed and radiated. A planet that absorbs more energy than it radiates should heat up, and vice versa (assuming no internal energy sources). The only significant external source of radiant energy for the Earth is the Sun, and this incoming sunlight is either absorbed by the Earth or reflected back into space. The **albedo** – the fraction of incoming rays reflected – depends on the surface of the planet: clouds, ice, and snow have relatively high albedos, while the values for dry land and water are lower. A watery planet that cools so much that its surface becomes covered with ice and snow will then reflect even more sunlight and cool even faster – the **runaway icebox** effect. Lacking liquid water on its surface, such a planet would not be a good prospect for the development of life.

This scenario was a distinct possibility for the early Earth because the early Sun was about 30 percent less luminous than it is today. As such, it would not have produced sufficient radiation to prevent the Earth from freezing over except for another important mechanism called the **greenhouse effect**, which uses the atmosphere to help warm the Earth's surface. In the greenhouse effect, visible light from the Sun passes through the Earth's atmosphere and is absorbed by the surface, which then reradiates energy back into space. But due to its much lower temperature compared to the Sun, the Earth's surface emits mostly infrared rays, rather than visible light. These infrared rays are then absorbed by certain molecules in the atmosphere – the so-called greenhouse gases, such as carbon dioxide, water vapor, methane, etc. – preventing the heat from escaping into space and keeping the lower atmosphere and surface warmer than normal. With carbon dioxide as one of its major constituents, the early atmosphere was an effective insulator, and the resulting greenhouse effect reduced the Earth's cooling rate and kept its oceans from freezing.

At some point during this first eon, life began on Earth. Unfortunately, the details are quite sketchy on this subject, but for very good reasons. First, the initial single-celled life forms were microscopically small, making any fossils of individual organisms extremely difficult to detect. Second, these initial organisms would not yet have developed hard body parts – bones, shells, teeth, etc. – which are more likely to be preserved as fossils than soft cells. And third, because the surface of the Earth was under heavy bombardment by asteroids and comets during most of this period, very few potential fossil-bearing rocks from this epoch have survived to this day.

The precise timing of this important event is thus uncertain; in fact, the formation of life on Earth might have occurred *numerous* times if the arrival of impacting bodies delivered enough energy to vaporize – or at least sterilize – the early oceans repeatedly. Major impacts continued on the Moon – and probably on the Earth – until about 3.8 billion years ago. With the earliest fossils dated at about 3.5 billion years ago, it seems likely that life arose for good by some time between 4.0 and 3.5 billion years ago. Whether it required 100 million years or only 10,000 years to produce living cells from collections of organic molecules is not yet known, nor is the actual sequence of the synthesis of the first DNA molecules and other constituents of the first cells.

The first life forms were very simple, compared to our present bodies. They consisted of single-celled **prokaryotes** – cells with no special structures inside the cell walls, such as a nucleus to contain the cell's DNA. Prokaryotes reproduce asexually by simple cell division, a process that generates identical copies of each cell, unless changes in the genetic code produce mutations. The variation that results from this method of reproduction is random and not particularly efficient in promoting evolution. And because the process of evolution has no special agenda to produce complex life, the earliest life forms were very slow to develop additional complexity. Life remained quite simple for hundreds of millions of years after it originated.

Just as we are uncertain of the timing of life's origin, we are also unsure of the setting for the first life. As water is the basis for terrestrial life, it is likely that a watery environment played a major role in the formation of the first living beings. In addition to water, the requirements for life usually include a supply of organic molecules and a source of energy. Organic molecules are needed for use as biochemical building blocks, from which the larger, more complex molecules can be fashioned. Energy is needed to initiate some **exothermic** (energy-releasing) reactions and to sustain **endothermic** (energy-absorbing) reactions.

One possible site for the start of life is in shallow tidal pools on the edges of landmasses. In these small bodies of water, dried by exposure to sunlight, sparked by lightning discharges, and irradiated by ultraviolet rays, the raw materials of life might have been concentrated and energized sufficiently to manufacture the principal molecules found in our cells today. Laboratory experiments – most notably by Stanley Miller in the 1950s – have demonstrated that a highly reducing atmospheric mixture of methane, ammonia, and water vapor that is exposed to electric discharges over a short period of time, can produce an array of interesting organic molecules, including amino acids. A less reducing – but perhaps more probable – atmosphere of carbon dioxide, nitrogen and water vapor proves to be much less efficient at producing organic molecules.

Others point out that life could have arisen at the bottom of the sea, where deep-sea vents provide geothermal energy for reactions, and raw materials would have been readily available to synthesize complex organic molecules. We now have evidence of such ecological systems on the ocean floor but cannot yet say for certain whether life originated there, in the tidal pools, or somewhere else. It is apparent, however, that by the end of the first eon, life was well established in the oceans of Earth, which covered the planet without a serious challenge from dry land.

The Second Eon: 3.6–2.6 billion years ago

By the Earth's second billion years of existence, the heavy bombardment phase was over, with only occasional impacts by asteroids and comets delivering matter and energy to the surface. Without a high frequency of impacting bodies and their supply of extraterrestrial energy, the Earth's surface was heated primarily by incoming solar radiation and convective heat flow from the interior while being cooled by its own emissions to space. The gradual cooling of the outer layers of the planet made the mantle and crust less flexible and served to restrict the convective process to the more plastic portions of the mantle, resulting in a limited number of hot spots where rising magma produced new sea floor or volcanic islands.

The crust, which became cooler and more rigid with time, eventually was broken into numerous crustal plates; these were then carried across the surface by the motion of the underlying mantle. Where the hot spots produced new crustal material, the plates were forced apart; where plates collided at subduction zones, the denser rock was plowed back into the mantle to be heated and reprocessed while the less dense rock was uplifted, eventually forming the beginnings of continents. These early continents remained relatively small; radioactive heating maintained the interior at a higher temperature than at present, and the greater convective heat flow that resulted prevented the formation of large plates.

As plate tectonics developed during this period, the continental landmasses gradually increased in size. Rocks that made up portions of these early continents can still be identified in some of our present continents. As more crustal rock rose above the level of the seas, more of it was eroded by the action of wind and water. Particles from these weathered rocks were carried to the oceans where they formed the basis of the first sedimentary rocks.

In addition to changes in the outward appearance of Earth caused by the growth of continents, the second eon also saw important modifications to the atmospheric composition. The principal atmospheric gases at the beginning of this period would likely have included diatomic nitrogen and carbon dioxide, with water vapor in a temperature-dependent equilibrium with the liquid water in the oceans. Carbon dioxide is soluble in water and would have dissolved in the oceans to some extent to form a weak solution of carbonic acid (H_2CO_3). This would have reacted with calcium ions – washed into the sea by weathering of the relatively new continental rocks – to form calcium carbonate, an insoluble precipitate. Deposition of this material on the ocean floors produced the first limestone beds by about 2.8 billion years ago.

These chemical steps thus provided a mechanism for the continual transfer of carbon dioxide from the atmosphere to the oceans to the rocks. Because carbon dioxide is an effective greenhouse gas, this gradual reduction in its atmospheric concentration also served to diminish the greenhouse effect. Normally, this would have allowed the planet to cool more efficiently, thus reducing the Earth's surface temperature and possibly allowing the oceans to freeze. However, the steady increase in the Sun's luminosity – which began with the solar system's formation – countered the decreasing greenhouse effect and kept the Earth's surface temperature high enough to maintain liquid water.

During this period, the Earth's oceans initiated another significant change in the atmospheric composition by hosting considerable numbers of single-celled bacteria, some of which had evolved to become **cyanobacteria** (or blue-green algae). Over time, these particular prokaryotes had developed the ability to utilize some of the most abundant molecules on the surface of the planet (carbon dioxide and water) in conjunction with sunlight to produce sugars – a form of stored energy. This reaction, known as **photosynthesis**, can be written as follows:

$$6CO_2 + 6H_2O + Light \rightarrow C_6H_{12}O_6 + 6O_2 \, .$$

It will be noted immediately that an important byproduct of this reaction is molecular oxygen, a molecule that has received little mention in our discussion of the Earth's early atmosphere. This is because oxygen in this form is highly reactive, combining readily with many elements to form a variety of stable oxides. The burning of wood in a forest fire sparked by lightning is one example of such a reaction; another less dramatic one is the gradual rusting of iron materials exposed to air. Because of its reactivity, oxygen normally winds up in combination with other elements; thus, without some ongoing production, free oxygen is bound to be rare.

A lack of free oxygen was actually good for early life forms, which had no experience living in an oxygen-rich environment and lacked the defenses needed to cope with this harsh chemical. (Some of these anaerobic bacteria still exist today, but they cannot survive exposure to air and the oxygen it now contains and thus must remain secreted away in safe locations, such as the intestinal tracts of humans.) With an early atmosphere devoid of this poisonous gas, these simple life forms flourished and performed evolutionary experiments in biochemistry, some of which led to the development of photosynthesis.

We might marvel that some bacteria should have successfully evolved the ability to generate a chemical that was toxic to their own species, but in fact this is not really that shocking. Photosynthesis was developed as a means of storing the Sun's energy, a process that is beneficial to the organism that can perform this feat. The fact that some toxic waste (oxygen) is produced is not a problem for the organism as long as this waste can be quickly dispersed into or absorbed by the organism's environment. (For most of us, garbage is not a problem – unless the trash collectors go on strike.) The first oxygen molecules produced by photosynthesis did not make a dramatic contribution to the composition of the atmosphere and therefore did not threaten the bacteria that produced them. Iron and other elements brought to the surface by volcanic action acted as oxygen sinks; they were readily oxidized, absorbing what little free oxygen had accumulated in the atmosphere and keeping its concentration low. Highly reactive oxygen could not become a major atmospheric component until the surface elements had become largely oxidized; only then would life have to ooze for cover and develop ways of coping with the poison.

During the second eon of planetary evolution, the Earth cleared a few more hurdles on its way to becoming a habitat for intelligent beings. Plate tectonics became firmly established and starter continents began to form. The oceans provided a mechanism to remove carbon dioxide from the atmosphere and store it in surface rocks, leaving nitrogen as the primary atmospheric component. The resultant lowering of the greenhouse effect countered the increasing solar radiation and maintained a relatively stable surface temperature. Life had not developed anything more complex than single-celled bacteria, but some of these had discovered a means to store the energy from sunlight, generating free oxygen in the bargain. The impact of this discovery was not to be felt until the next billion years, but it was to revolutionize the course of future evolution.

The Third Eon: 2.6–1.6 billion years ago

During the Earth's third eon the processes begun in earlier eras continued to operate. Heating of the interior by radioactive decay diminished as the supply of parent isotopes was gradually reduced. Decreased heat flow requirements allowed the mantle convection rate to decline, and the less vigorous forces acting on the crustal rocks would have promoted a surface with fewer crustal plates and permitted the formation of larger continents. By about 2 billion years ago, plate tectonics was probably functioning much as we observe today, although the continental pattern would have been quite different from that of the present Earth.

Occasional impacts of asteroids and comets would undoubtedly still have occurred, and at a rate higher than we observe today. However, as the rock record from these early periods is somewhat sparse, direct evidence of individual events is essentially nonexistent. The effect of these impacts on the existing life forms is also difficult to ascertain, but by then the microbes had already managed to live for over a billion years under similar conditions of bombardment, and they were probably not greatly hindered by such catastrophes.

The most notable development in this era was the rise of the free oxygen supply in the atmosphere, due to photosynthesis carried out by the cyanobacteria. These organisms had already been on the planet for a billion years or so, but the Earth's surface conditions had not yet been optimized for their needs. However, the growth of the continental landmasses at this time provided an increase in the surface area of the shallow seas, where these water-dwelling bacteria found adequate sunlight for practicing their craft. At first, the oxygen produced by the cyanobacteria reacted readily with other elements and molecules in the sea, preventing any significant buildup of free oxygen in the air. Only as the elements in the oceans became oxidized did the unreacted oxygen begin to accumulate in the atmosphere. Evidence of this atmospheric change comes in the form of the **banded iron formations** – alternating layers of reddish, iron-rich layers and gray, iron-poor layers of sedimentary rock. Iron in its reduced state – delivered to the sea by erosion and volcanic activity – reacted with oxygen to produce insoluble iron oxides, which precipitated to the seafloor to form sedimentary layers with the reddish color of rust. The gray layers (which lack iron oxides) could have resulted from an under-abundance of either the iron or the dissolved oxygen, possibly due to sporadic volcanic activity or variations in the production of oxygen by the bacteria. (One hypothesis is that the colonies of cyanobacteria expanded so fast that their production of the toxic oxygen waste product outstripped the supply of iron in the local environment, causing mass extinction of the bacteria by oxygen poisoning. Only after the remaining bacteria had recovered from this disaster and begun to generate oxygen could the iron oxide bands form once again.)

The banded iron formations begin to appear in rocks dated from 3.5 billion to 2.5 billion years ago; their frequency increases to a peak about 2.5 billion years ago and diminishes until they stopped forming about 1.8 billion years ago. Before the evolution of cyanobacteria there was no significant source of free oxygen, and iron remained in its reduced state. But as the number of these bacteria increased, the free oxygen they produced converted the iron to its insoluble, oxidized state and made more of these formations possible. Following the peak, the supply of iron remaining in the reduced state dwindled, and oxygen began to accumulate in the atmosphere. Since 1.8 billion years ago, the iron at the surface of the planet has been essentially completely oxidized and thus insoluble in seawater. With no reduced iron available to be dissolved in the sea, none could be oxidized and precipitated as new sediment, and the banded iron formations could no longer be created.

It appears fairly clear that during this third eon the atmosphere's oxygen content was increasing while the carbon dioxide content was decreasing (due to photosynthesis and the formation of limestone as previously discussed). The result was that approximately 1.8 to 2.0 billion years ago the Earth switched over from a nitrogen/carbon dioxide atmosphere to a nitrogen/oxygen atmosphere. This change had major implications for the future evolution of life.

Creation of free oxygen in the atmosphere made possible the formation of another key atmospheric molecule – ozone (O_3), which is produced in the upper atmosphere in a two-step process. First, molecules of oxygen (O_2) are dissociated into *atoms* of oxygen by ultraviolet rays (O_2 + sunlight $\rightarrow O + O$); then the atomic oxygen combines with molecular oxygen to form ozone ($O + O_2 \rightarrow O_3$). The ozone molecule in turn can absorb somewhat less energetic ultraviolet radiation and split apart again (O_3 + sunlight $\rightarrow O_2 + O$) with the oxygen atoms then combining with ozone to form molecular oxygen ($O + O_3 \rightarrow 2O_2$). By this cyclic process, absorption of ultraviolet rays by both O_2 and O_3 converts solar energy into thermal energy, thus raising the air temperature. Without ozone, a greater proportion of ultraviolet rays would penetrate the atmosphere, exposing the surface of the Earth and the organisms on it to potentially harmful radiation. The introduction of ozone would eventually make it much easier for life to emerge onto dry land without first having to develop protection against the Sun's ultraviolet rays.

The development of an oxygen-rich atmosphere midway through this period changed the ground rules for the existing microbes. With this toxic oxygen waste from the cyanobacteria now spiraling out of control, the microbes were faced with only a few choices: (1) they could run and hide, seeking out spaces where the poisonous oxygen did not penetrate, and go on living as they always had; (2) they could develop some sort of protection against this reactive substance while finding ways to utilize oxygen for their own good; or (3) they could simply perish.

Many chose the first route. There are a host of anaerobic bacteria that live quite successfully today in locations where they avoid contact with air; they cause infections inside the body and are responsible for peritonitis, appendicitis, botulism, tetanus, and gangrene. There were other such bacteria that apparently became extinct, having disappeared from the fossil record around the time of the atmospheric transition. Some bacteria developed the means to use oxygen to obtain energy through **aerobic respiration**: oxygen + organic matter → carbon dioxide + water + energy. And some organisms teamed up with others to create an entirely new type of cell that could function efficiently in the new oxygen-rich environment: the **eukaryote**.

Eukaryotes differ from prokaryotes in several ways. Eukaryotes are generally larger – 10 to 100 microns in diameter compared to 0.2 to 10 microns for the prokaryotes (one micron = 0.001 millimeter). Genetic material in the eukaryote is organized inside a nucleus that is bounded by a membrane, while in the prokaryote (which lacks a nucleus) it is in a single strand of DNA inside the cell. Most eukaryotes contain organelles – small bodies inside the cell, such as mitochondria and chloroplasts, which perform specialized functions; prokaryotes have no organelles. Most eukaryotes employ an aerobic metabolism, allowing them to thrive in an oxygen-rich atmosphere while prokaryotes as a group utilize a variety of anaerobic and aerobic processes. Eukaryotes are important to us because they make up all of what we regard as *complex* life on Earth, including ourselves.

The timing of the evolution of the first eukaryotes is still unclear; these tiny, single-celled organisms with soft bodies were not very conscientious about leaving fossils to illustrate their growth and development. However, with the observed difference in size between present-day prokaryotes and eukaryotes, the cell types of certain microfossils can be inferred from their relative sizes. Studies of the size and distribution of fossils of spheroidal micro-algae imply that significantly larger cells – which may well have been eukaryotes – evolved by about 1.8 billion years ago, around the same time when the transition to an oxygen-rich atmosphere occurred. More definitive evidence of the existence of eukaryotes appears later in the fossil record, but it does not seem unreasonable that these aerobic species would have evolved at the time when free oxygen became abundant on the planet.

The *process* by which eukaryotes formed is also somewhat mysterious, but there is good evidence for organelles such as the mitochondria and the chloroplasts originating as independent prokaryotes, performing aerobic respiration and aerobic photosynthesis, respectively. In some instances, envelopment and assimilation of these organisms by larger cells resulted in a cooperative arrangement in which the new organelles provided services to their host cells in exchange for a supply of nutrients. The mitochondria took on the job of using oxygen to perform respiration for the cell, releasing energy from organic matter; those cells equipped with chloroplasts were able to *store* solar energy in molecular bonds, producing oxygen as a byproduct of photosynthesis. This symbiotic relationship proved to be an excellent way for organisms to survive in an oxygen-rich environment; with a few modifications, the eukaryotes would be poised to launch a vigorous expansion and diversification program in the next eon.

The Fourth Eon: 1.6–0.6 billion years ago

The fourth eon saw no new major processes introduced that would make dramatic changes in the Earth's surface and atmosphere. Instead, the trends established by existing processes during the previous billion years of evolution were maintained, guiding the surface environment gradually closer to that of the present.

Plate tectonics continued to operate as before, but with ever-increasing amounts of lower density rock protruding above the global oceans. The continental pattern – which still bore essentially no resemblance to our current arrangement – was constantly modified as continental masses were split apart, rotated, and transported thousands of kilometers over the surface to collide again with each other at some time in the future. New seafloor was still being created at mid-oceanic rifts while denser crustal material was recycled and re-melted where plates met at subduction zones. Changes in elevation of the various continental masses and/or changes in the level of the seas controlled the opportunities for deposition of sediment eroded from the exposed continental rocks; portions of the currently exposed crustal rocks dating from this era can be found on several of the present continents.

As before, occasional impacts by comets and asteroids undoubtedly occurred, although we have no direct evidence of individual events. The impact rate would have been considerably less than it had been in the first eon as most of the larger meteoroids had been gradually swept up by the inner planets whose orbits they intersected. Effects of these collisions on the simple life forms that existed then on the planet are difficult to ascertain, but apparently none were violent enough to provide any serious setback to our evolving ancestors.

The major developments during this period included several important evolutionary changes in the existing life forms. Those life forms most involved in these changes were the eukaryotes, which had just developed in the previous eon. The prokaryotes, on the other hand, continued to go on about their business without making any significant modifications.

By around 1.2 billion years ago, some of the eukaryotes had evolved multi-cellular life forms consisting of different types of cells with special functions, all of which worked together to form larger, more complex organisms than had previously existed. No longer would each individual cell need to interact directly with the environment in order to live, nor would it have to be able to survive independently. The reality of multi-cellular life opened the door to a nearly unlimited variety of potential life forms, including humans.

By about 1.1 billion years ago the eukaryotes had developed specialized cells for reproduction and invented **sexual reproduction**. The early eukaryotes reproduced by **mitosis** – a process in which the cell divides and its DNA creates a copy of itself. Daughter cells produced in this manner are essentially identical to their parent cells, aside from random mutations. On the other hand, sexual reproduction by **meiosis** involves two different cells (egg and sperm), each contributing half the DNA of the new daughter cell, ensuring that the daughter will be different from either parent. By forcing each generation to be slightly different from the previous one, sexual reproduction increases the diversification within each species and thus accelerates the process of natural selection.

By the end of the fourth eon a somewhat different type of organism had evolved – a eukaryote that obtained its carbon from organic matter, rather than from carbon dioxide. These forerunners of today's animals existed as predators, consuming other cells and utilizing the organic molecules within them. The world that had previously been populated by **autotrophs** – plant-like organisms capable of using carbon dioxide as their sole carbon source – now became the home of a variety of **heterotrophs** – animal-like organisms capable of using organic matter as their sole carbon source. The rise of heterotrophs on the planet meant that some of the population of autotrophs was destined to become the basis for a food chain. The exact timing of this development is difficult to determine as the early animals had no bones, shells, or other hard parts to form obvious fossils, but metazoans – multi-celled animals – appear in the fossil record about 600 million years ago.

During all this time, the cyanobacteria continued to process carbon dioxide from the atmosphere into free oxygen, increasing the latter while decreasing the former. Reduction of atmospheric carbon dioxide in turn diminished the greenhouse effect and allowed the Earth's surface to cool. The cooling was so extensive that a series of ice ages occurred between 800 and 600 million years ago, as evidenced by glacial deposits in the rock layer. Such a dramatic change in temperature over the globe must have provided a significant stress to the microbes and probably slowed the assault of the cyanobacteria on the atmospheric carbon dioxide.

Another check on the activity of these enterprising bacteria at this time was the rise of predatory microbes – the heterotrophs. Previous forms of life had managed to perform chemical reactions to extract nutrients and energy from the environment. But animals cannot do this and instead must ingest organic matter in order to live. The cyanobacteria declined in numbers around 600 to 700 million years ago, probably due to the action of their newly developed predators. Without the invention of animals at this time to slow the depletion of atmospheric carbon dioxide by the cyanobacteria, the ice age might have lasted considerably longer, making further evolution of life much more difficult.

The Current Eon: 600 million years ago to the present

Up to this point, the Earth had managed to produce a variety of small, relatively simple, life forms. Most of them were single-celled organisms, but a few had begun to evolve into more complex, multi-cellular structures capable of sexual reproduction. This biological groundwork, together with the Earth's contributions of an oxygen-rich atmosphere (for efficient energy production through aerobic respiration), significant continental landmasses (providing a wide range of habitats for evolving life forms), and established plate tectonics (supplying a driving force for gradual changes in the Earth's surface), made possible a revolution in the diversity of life on the planet – now called the **Cambrian explosion**.

During this relatively short period of time – from about 570 to 530 million years ago – the lineages of most of the major divisions of the animal kingdom were established. This is not to say that *all* of the different species we find today on Earth – cats, jellyfish, cockroaches, earthworms, sponges, etc. – were developed at this time; rather, the early ancestors of these species had differentiated themselves from one another using characteristics that still differentiate today's modern species.

Once this was done, the stage was set for an acceleration of the evolutionary process as the newly developed animals diversified to fill the host of environmental niches for which no creatures had previously existed. The result is that the vast majority of the different creatures and plants we find around us today came about comparatively recently, developing in the last 500 million years or so – the last 12% of the Earth's history. Plate tectonics has helped to drive this diversity; the latest round of continental drift, which produced the continental structure that we observe today, has proceeded over the last 200 million years.

The path of evolution was not smooth or straightforward. In fact there were a number of dramatic events that occurred on the planet – such as impacts by asteroids or comets – that resulted in **mass extinctions**. These events completely eliminated large numbers of species in a relatively short period of time. The five events that wiped out at least 70% of existing species occurred about 438, 367, 250, 202, and 65 million years ago (Powell 1998).

The dinosaurs perished in the most recent of these events, apparently as the result of an asteroid impact on the Yucatan peninsula about 65 million years ago. The dinosaurs were not all crushed by the descending asteroid – which was probably about 175 kilometers in diameter. Instead, they would have died because the impact produced clouds of dust in the atmosphere that screened off sunlight and persisted for many years, destroying the vegetation that was the ultimate food source for the dinosaurs. The dinosaurs – along with 70% of all species on Earth at the time – died out because they could not adapt to survive under the modified conditions produced by the impact. While this impact was disaster for the dinosaurs, it was an opportunity for the mammals, which were still in their early stages of development. By clearing the dinosaurs out of the way, the asteroid impact left numerous ecological niches wide open, to be filled by the surviving species and their descendants – including humans. Therefore, it could be said that we exist *because* of this impact, not in spite of it.

The dinosaurs existed as the dominant life form on Earth for about 150 million years before they were toppled from power by an impact; humans have been around for far less time than that. Could our species be destroyed by another impact, thus making way for another intelligent species to evolve on Earth? Probably – it does not take a really big asteroid to ruin a civilization, and there are plenty of asteroids around. Would another intelligent species actually evolve to take our place? Perhaps, but we do not really know the probability of the evolution of intelligent species on a planet that already has life. That is another factor in the Drake equation.

The evolutionary path to intelligence is not mapped out in advance. No particular species in existence a few hundred million years ago was predestined to evolve into today's intelligent beings, but there was a species that did just that, and we may be able to trace our lineage backwards in time to discover it. In a similar sense, while it is possible to research your ancestry and discover information about your great-great-great-great grandparents, it is quite impossible to make detailed, accurate predictions about your own great-great-great-great grandchildren, who may or may not eventually exist. We can say that the odds of one of *your* descendants becoming President or traveling to Pluto – whether you desire this or not – will be higher if you have *lots* of descendants. But if you die young, before you reproduce, then those odds fall to zero.

It should be clear by now that the Earth is a very complex place, with an equally complex evolutionary history. The next questions that naturally arise concern the uniqueness of our planet and its development of life. How similar to Earth must planets be in order to be capable of evolving life? How abundant are such planets in our Galaxy? The next chapter will provide a brief overview of what we know about other planets – both in our solar system and around other stars – in hopes of arriving at answers to these questions.

MAIN IDEAS

- The present Earth has a number of properties that make it unique among the planets with which we are familiar, but we do not know exactly which features have been essential to the evolution of intelligent life here.

- The high abundance of carbon in terrestrial life – despite its relatively low abundance on Earth – provides a strong argument that if extraterrestrial life exists, it will most likely be based on whatever carbon is accessible in its environment.

- Radioactive decay and sedimentary rock layers provide mechanisms for determining the ages of rocks and the fossils contained within them.

- The fossil record can be interpreted to trace the evolution of life on Earth, but because this record is necessarily incomplete, it does not always reveal the detailed steps by which evolution has proceeded.

- Evolution of intelligent life on Earth required the coordinated development of Earth's surface and atmosphere over a time scale of several billion years; extraterrestrial intelligent life is unlikely to evolve on a significantly shorter time scale.

KEYWORDS

abundance
atmosphere
Cambrian explosion
core
Darwin, Charles
endothermic reactions
fossils
half-life
isotopes
mantle
metamorphic rock
parent isotope
planetesimal
punctuated equilibrium
runaway icebox
solar wind
Van Allen radiation belts

aerobic respiration
autotrophs
continental drift
crust
daughter isotope
eukaryote
gametes
heterotrophs
Jovian planets
mass extinctions
mitosis
photosynthesis
plate tectonics
radioactive decay
sedimentary rock
stasis

albedo
banded iron formations
convection
cyanobacteria
DNA
exothermic reactions
greenhouse effect
igneous rock
magnetosphere
meiosis
natural selection
phyletic gradualism
prokaryotes
reducing atmosphere
sexual reproduction
terrestrial planets

LAUNCHPADS

1. If aliens do exist on other planets, which of the Earth's current physical characteristics are such planets most likely to share?

2. Should extraterrestrial civilizations have evolved more rapidly, more slowly, or at about the same rate as our terrestrial civilization?

3. Suppose that the dinosaurs had not been wiped out by an asteroid impact 65 million years ago. Would we be here now?

REFERENCES

Baugher, Joseph F. 1985. *On civilized stars: The search for intelligent life in outer space.* Englewood Cliffs, NJ: Prentice-Hall, Inc.

Cox, Arthur N. ed. 2000. *Allen's astrophysical quantities.* 4th ed. New York: Springer-Verlag.

Darwin, Charles. 1859. *The origin of species by means of natural selection, or the preservation of favored races in the struggle for life.* New York: Modern Library.

Eldredge, Niles, and Stephen J. Gould. 1972. "Punctuated equilibrium": An alternative to phyletic gradualism. In *Models in paleobiology*, ed. T. J. M. Schopf, 82-115. San Francisco: Freeman, Cooper and Co.

Emsley, John. 1999. *The elements.* 3rd ed. New York: Oxford University Press.

Lang, Kenneth R. 1980. *Astrophysical formulae.* 2nd ed. New York: Springer-Verlag.

Lide, David R. ed. 1998. *CRC handbook of chemistry and physics.* 79th ed. New York: CRC Press.

Powell, James L. 1998. *Night comes to the cretaceous: Comets, craters, controversy, and the last days of the dinosaurs.* San Diego: Harcourt Brace.

Chapter 5
OTHER PLANETARY ENVIRONMENTS

In which the layout of our solar system is described, the physical properties of other solar system bodies are presented, expectations for extraterrestrial life in the solar system are discussed, techniques used in searching for planets around other stars are explained, and our current knowledge of extrasolar planets is reviewed.

We live on the Earth – a planet – one of several bodies of decent size orbiting the Sun. Planets seem to be reasonable places to evolve life: they provide locations where the elements of life can be concentrated and nurtured at appropriate temperatures over long periods of time. They also offer a variety of environments – liquid oceans, gaseous atmospheres, and/or solid surfaces – for life to utilize as it develops. Planets may not be the *only* places where life can evolve, but because our sole example of a biological system exists – and thrives – on a planet, we should certainly make planets the principal target of our search. If our solar system is representative, there are several different types of planets, and they will not all be equally hospitable to living beings. The abundance of extraterrestrial civilizations is thus linked to the abundance of suitable planets. In the previous chapter we attempted to identify which planetary properties were necessary for the formation and evolution of terrestrial life; in this chapter we will discuss how common planets with these properties may be within our Galaxy. Additionally, we will investigate the techniques used to search for planets *outside* our solar system and report on progress made to date.

TERMINOLOGY

Planets are bodies that orbit stars, as do asteroids, comets, and other stars. The distinction among these various objects is not always clear, but it is generally connected with mass and/or composition. Stars (which will be discussed in detail in the next chapter) are the most massive of this group, with masses so great that their intense gravitational fields compress and heat the stellar gases to nuclear ignition temperatures. The nuclear energy released inside a star maintains its surface at such a high temperature that the star emits visible light. Stars can be found orbiting each other, in systems called **binary stars**.

The masses of planets are considerably smaller than those of stars, resulting in less intense gravitational fields and lower temperatures. Planets are too cool to generate nuclear energy or radiate visible light, but they do give off infrared rays; thus, although they do not shine as stars do, they *are* sources of invisible radiant energy just as people are. And just as we see each other by the light that reflects off our bodies, we can observe the planets in our own solar system by the sunlight they reflect off their surfaces. The lower temperatures found on planets also permit the matter there to be liquid and/or solid rather than completely gaseous as in stars. With its much greater mass, the star in a planetary system tends to remain relatively stationary while the planets do most of the moving as they orbit about it.

The lower limit on stellar masses is around 8% of the mass of the Sun (or about 80 Jupiter masses); bodies less massive than this cannot sustain nuclear reactions to supply their own energy.

This *could* also be taken as the upper limit on planetary masses, but with no observable bodies in our solar system in this mass range, astronomers have been uncomfortable letting 'planets' be this large. Instead, based on models of the interior structures of such objects, astronomers have set the upper limit on planets at about 10 to 13 Jupiter masses and covered the range of objects from 13 to 80 Jupiter masses with the label '**brown dwarf**'. With their low temperatures and reduced light outputs, brown dwarfs are very difficult to locate and study, and only a few good candidates have been identified. However, it is estimated that there may be up to 10% as many brown dwarfs as there are stars – perhaps 10 to 40 billion of them (Basri 2005).

Asteroids and comets are much less massive than most of the planets in our solar system. However, there are no clear definitions of the minimum mass of a planet or the maximum mass of an asteroid or comet. This is because our discoveries of these objects have been spread out over several centuries, with the terminology invented and modified over the same period of time as new objects were found. The distinctive features of asteroids and comets will be discussed later in this chapter.

PLANET FORMATION

As mentioned above, planets orbit stars – at least, the planets in our solar system orbit the Sun. This fact is linked to the manner in which planets and stars are formed. Stars form out of huge clouds of dust and gas (**nebulae**), which can be observed at numerous locations along the Milky Way. (Figure 1.6 showed two such locations, in the constellations Orion and Sagittarius.) Given the right conditions, gravity can pull such a cloud together, concentrating its matter toward the center of the cloud and heating it until it becomes a star. However, this gravitational contraction is generally not able to collect *all* of the matter from the cloud into the forming star before the nuclear reactions begin, and it is this leftover matter that forms into planets, asteroids, and comets, which orbit outside the star. Thus, planets are byproducts of star formation.

Do *all* stars produce planets as they form? This is unknown as yet. We have detected evidence of planets around *some* other stars and therefore know that the Sun is not unique in that respect. But there may well be stars that have no planets; for example, many binary star systems may prevent the formation of planets in stable orbits about them. Are there any planet-sized bodies that do *not* orbit any star but instead drift through space alone? While such planets are not likely to form independently out of clouds of gas and dust, it is possible that planets can be ejected from a planetary system by close interactions between stars or planets. If such planets do exist, they will be quite difficult to locate and study, due to their lack of a nearby star. For the same reason, they will probably also make poor homes for extraterrestrial civilizations and thus will not be considered here.

OUR SOLAR SYSTEM

Our study of planets will rely heavily on those closest to us – the members of our solar system. These planets have been observed for centuries by Earthbound astronomers and examined more closely by spacecraft over the last few decades. We have become fairly familiar with the Sun's planets – Mercury, Venus, Earth, Mars, Jupiter, Saturn, Uranus, and Neptune (Pluto was reclassified as a **dwarf planet** in 2006) – knowing them as physical bodies with individual surface features rather than just points of light in the night sky. They will form the basis of comparison for any extrasolar planets we may find or imagine.

The first step in understanding our solar system is to gain some perspective on size and distance. This can best be done by constructing a scale model of the solar system – one in which the sphere used to represent each planet is the correct relative size with respect to the others. The first planet chosen determines the scale for the whole model, and the other planets follow from it.

For this model, we will begin with our home planet because it is most familiar to us. Let the Earth be represented by a marble, just over a half inch in diameter. On this scale, the Moon is a BB, about four millimeters across. Venus, slightly smaller than the Earth, is another marble, about a half inch in diameter. Mercury, a five-millimeter-diameter steel ball bearing, is smaller than the Earth and Venus but larger than the Moon. Mars (the red planet) is a ball seven millimeters across, formed from a pink eraser. These inner planets are all represented by relatively small objects, but that is because there are bigger things to come in the model.

Beyond Mars, the planets get considerably larger. Jupiter is a plastic foam ball six inches in diameter, larger than any other planet in the solar system. Saturn is nearly as large, being a foam ball five inches across. Uranus is a smaller foam ball, only two inches across, and Neptune is the same thing but slightly smaller. These four model planets appear distinctly different from the first four, both in size and in composition; this difference will be further explored very shortly.

The last, most distant body in our model is Pluto; it is also the smallest, being a very tiny, shiny ball only three millimeters across. Pluto is even smaller than the Moon, which only adds to the problem of distinguishing classes of objects by size. (Pluto's reclassification from 'planet' to 'dwarf planet' was based partly on its size and partly on its orbital properties.) There are many other objects in the solar system that could be added to the model, but most of them would be too small to see at this scale. Figure 5.1 depicts the relative sizes of the model solar system bodies introduced so far.

One object that we certainly cannot overlook is the Sun, represented by a sphere five feet in diameter (or a circle on the blackboard of similar size). The Sun is obviously the largest body in the solar system with ten times the diameter of Jupiter and over 100 times the diameter of Earth. The Sun's great mass (333,000 Earth masses) makes it the center of the solar system and produces a strong gravitational field that holds the planets in their orbits.

This scale model clearly demonstrates the relative sizes of the planets, Moon, and Sun, but it can also show the relative spacing of these objects throughout the solar system. Using the same scale, we can position the different objects in the model according to the sizes of their orbits; for example, the Moon BB should be located about 17 inches from the Earth marble, which in turn would be found about 180 yards from the blackboard Sun. The other planets would also orbit the Sun, at distances ranging from Mercury's 70 yards to Jupiter's half mile to dwarf planet Pluto's four miles. (This should explain why we knew little of the outer planets' surfaces before they were visited by our space probes.) The whole model solar system would be strung out over a planar disk about eight miles across, with the Sun at the center and the various tiny planets revolving about it – hardly an impressive sight when viewed from afar.

From how far away might extraterrestrials be viewing us? The closest they could be living would be the nearest Sun-like star to us, Alpha Centauri. On our scale model, Alpha Centauri would be another five-foot-diameter sphere located 27,000 miles away from the solar system. From this distance, the model Sun would appear as a mere point of light and the model planets would be quite invisible, shining only by the small amount of sunlight they reflect, and lost in the glare of the Sun. Indeed, we do not expect to *see* planets orbiting other stars, but we have other evidence that they exist.

Planetary Types

In looking over the model solar system, we see that the planets can be grouped together according to their appearance. The first four model planets are all fairly small and made mostly of glass or metal, while the next four are much larger spheres of plastic foam. We label these two groups for the prototype planet in each: Mercury, Venus, *Earth*, and Mars comprise the **terrestrial planets** while *Jupiter*, Saturn, Uranus, and Neptune are the **Jovian planets**. These groups provide a convenient way to discuss some general properties of the planets.

The planets within each group share common characteristics that are quite different from those of the other group; for example, terrestrial planets are small in size while Jovian planets are large. We can also note that terrestrial planets have low masses while Jovian masses are high, which should not be surprising, given their relative sizes.

The mass and size of a planet can be combined to yield the planet's average density (mass per unit volume), a measure of how closely packed the matter is within the planet. Here we find that terrestrial planets have relatively high densities while Jovian planets have low densities, as could be guessed from the model. The planet's density gives us a clue to its composition: the high-density terrestrial planets are mostly solid, made primarily of rocks and metals while the low-density Jovian planets are gaseous, composed of ices and gases.

The above features serve to sort the planets into the two groups, based on their sizes and compositions. Several additional properties – although not defining characteristics – reinforce the terrestrial/Jovian distinction. The model shows that terrestrial planets are close to the Sun while Jovian planets are farther away; because of this fact, the terrestrial planets revolve (orbit) rapidly about the Sun while the Jovian planets revolve slowly. Each planet also rotates (spins) on its axis, a motion that causes day and night on Earth. Perhaps surprisingly, the terrestrial planets tend to rotate slowly while the Jovian planets rotate rapidly. Most of the planets have moons in orbit about them; terrestrial planets have relatively few moons while Jovian planets have many moons.

In these examples of general properties of the two groups there are *no* exceptions; *every* terrestrial planet has lower mass, slower rotation, fewer moons, etc. than *any* Jovian planet. Each of the planets from Mercury out to Neptune lands squarely in either the terrestrial or Jovian camp; but one planetary object was not named in either group – where does Pluto belong?

Pluto is small in size (terrestrial) and low in mass (terrestrial), but its average density is low (Jovian). It is far from the Sun (Jovian) and revolves slowly (Jovian), but it rotates slowly (terrestrial) and has only a few known moons (terrestrial). Clearly Pluto does not belong in either group and has been classified as *neither* terrestrial nor Jovian; this was also a factor in Pluto's 2006 designation as a dwarf planet.

What is the reason for the solar system's division into terrestrial and Jovian planets? Will other planetary systems have terrestrial planets and Jovian planets, with the terrestrials close in to the star and the Jovians farther out?

We have plausible explanations for the layout of planets in our solar system. As mentioned above, the planets are believed to have formed from the solar nebula; as the Sun formed in the center, the planets condensed from the leftover cloud materials. Clearly, temperatures would have been higher toward the center of the cloud and lower farther out. The only materials that can condense into solid form at high temperatures are those composed of heavier elements – the metals and rocks. Because these elements are in relatively low abundance in the universe, the planets comprised primarily of heavier metals would be fairly small, with low masses. The low gravities of these planets would prevent them from retaining the hydrogen and helium atoms that were in highest abundance in the nebula, thus preventing such planets from growing very large.

Planets made in the cooler regions farther out in the cloud would be able to form from a wider variety of materials, including the gases and ices made from the lightweight elements such as hydrogen, helium, carbon, oxygen, and nitrogen. As these elements were very abundant in the nebula, much larger planets resulted, and because the planets were larger and more massive, their gravity would have been strong enough to hold more of the gaseous elements. This process, which produced terrestrial and Jovian planets in our solar system, should also have been operable in many – perhaps most – other planetary systems. What do we know about other planetary systems?

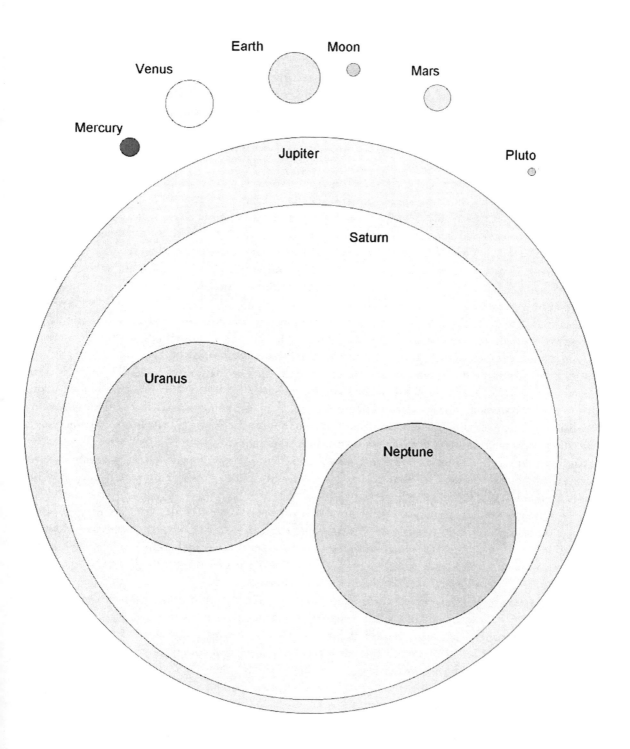

Figure 5.1: The model solar system (without the Sun – a 5-foot diameter circle).

EXTRASOLAR PLANETS

In recent years astronomers have discovered a number of these **extrasolar planets** (or **exoplanets**) orbiting relatively nearby stars. The discovery technique is not as simple as looking at a star through a telescope and watching for planets moving about it (as can be used to study the larger moons of Jupiter). With our current technology, planets are not directly observable at the great distances at which stars lie. Planets are relatively small and thus would appear much fainter than the stars whose light they reflect. And at its great distance from us, such a tiny planetary speck would seem to be so close to the star it orbits that it would be lost in the star's glare. Even so, there are some good options for detecting planets about other stars.

In any orbiting system in space, all bodies are free to move. For example, as the Earth orbits the Sun, the Earth does most of the moving, but the Sun wiggles just a bit in response. If no other planets were present, the Sun's center would describe a tiny orbit about the center of mass of the Earth-Sun system as the Earth traveled about its much larger orbit. If the Earth were a more massive planet, the Sun's corresponding motion would be larger and more detectable by distant observers. It is this reactionary wiggle of a star – in response to the much larger orbital motion of a much smaller and fainter planet – that astronomers have been detecting since the mid 1990s. (The actual detection involves changes in the spectrum of the star's light, rather than shifts in the star's position in the sky, details of which will not be discussed here.) The challenge is to use the observed motion of the star to deduce information about the planet (or planets) that caused it. But this challenge is being successfully met, with the majority of the dozens of extrasolar planets known having been discovered by this method.

Another technique being attempted is to measure the change that occurs in the *brightness* of a star when an orbiting planet passes between the star and the Earth. Such events have been observed within our own solar system; during a **transit** of Venus (or Mercury), the planet can be observed moving across the disk of the Sun. Because the disks of distant stars are not resolvable as yet, we cannot actually observe their planets in transit. But we can hope to detect the loss of light that should occur when a planet covers part of the star's disk – a measurement that should be easier for larger planets. So far this technique has been successful in detecting a handful of large planets orbiting closely about their parent stars. And in a few of *these* cases, a bonus measurement has been made.

As previously noted, planets are generally much smaller and cooler than the stars they orbit and thus are considerably fainter, making them nearly impossible to detect in the glare of the star. However, if conditions are right, the planet's radiation – or more properly, *loss* of radiation – can be measured in a system in which transits have already been detected. For if a planet is transiting its star, then half an orbit later the planet must pass *behind* the star in an eclipse.

Astronomers have measured eclipsing binary *stars* for decades, noting the drop in the combined light from both stars when one of them is eclipsed. The principle is the same for a planet-star system, except that the planet will usually be so much fainter than its star that an eclipse of the planet would normally pass unnoticed. However, if the star is relatively small and cool while the planet is relatively large and warm (although still smaller and cooler than the star), then the loss of the planet's radiation during an eclipse may reduce the total signal from the system by a measurable amount, perhaps a few tenths of a percent. And if transits have already been observed for a star, then astronomers will be alerted to watch for eclipses as well. Recently, astronomers have begun to measure such eclipses in a few of the transiting systems, obtaining better estimates of the planets' temperatures and ultimately, information on the compositions of their atmospheres. Improvements in telescopes and detectors should allow these methods to become commonplace in the future.

Still another detection plan would involve looking for the spectral signatures of key atmospheric molecules in the light reflected by an extrasolar planet. For example, the presence of molecular oxygen or ozone in a planetary atmosphere should be a good indication of life on the planet, as these highly reactive molecules require constant regeneration in order to build up any significant abundance. Although this approach is not currently feasible, it may become so in the future with better instrumentation.

Astronomers continue to get better at making and interpreting observations of extrasolar planets. To date, the number of extrasolar planets found – primarily by detecting the star's orbital motion – is well over one hundred. Most of these planets appear to be both large (several times the mass of Jupiter) and close to the star (one astronomical unit or less in many cases). Thus it would seem that these planetary systems contain large, probably gaseous, (and therefore Jovian) planets close in to the star. Does this observational fact negate our theories of planet formation? Does it mean that our solar system is abnormal in its arrangement – that perhaps our planetary system is unique and the only one to have life?

There are two points to be made here, the first being that a sampling of a few dozen planetary systems in a galaxy containing billions of stars is not necessarily going to be representative. We must be careful not to draw hasty conclusions from the initial data. Second, there is a very good reason why the searches so far have primarily found Jovian planets close in to their stars. The search technique being used is designed to detect motion of the star; this motion will be enhanced by either a larger planet or a closer planet (or both), which will result in greater gravitational attraction between the star and planet and more rapid motion of both objects (making the star's motion more easily detected). In short, the main objects discovered by the search are those that best match the search technique, not necessarily those that are most abundant; this is called a **selection effect**.

Selection effects are very common in astronomy and should be taken into consideration whenever possible, to avoid jumping to conclusions. In this case, it is quite reasonable to suspect that selection effects are involved and that terrestrial planets – as yet undetectable – may exist in some of the systems studied.

It would appear from the preliminary observations, however, that Jovian planets are not confined to the outer reaches of every planetary system, and this could have important consequences for our search. Some astronomers have proposed that Jovian planets should indeed form at greater distances from the central star, but that in some systems, they have migrated inward, interacting with other planets and perhaps disrupting the orbits of the terrestrial planets they encounter. In such cases, the system may no longer contain any terrestrial planets. Are these systems the norm or the exception?

There is another possibility that should be considered because it produces a rather different result. With the detection method being used, the planet's presence, mass, and orbital size are all deduced from the apparent motion of the star. However, in most cases, the best that can be obtained is a *lower limit* on the planet's mass. The actual planetary mass could be much greater, perhaps enough to change the designation of some of these bodies from 'planet' to 'brown dwarf'.

Brown dwarfs are massive enough that they may constitute binary systems with their central stars; as such, they may make stable planetary orbits difficult to obtain. This would mean that some of the 'planetary systems' that have been identified are actually binary systems in which the fainter component is a brown dwarf – a failed star. Such systems would probably be less likely to make good homes for planetary-based life.

ALSO-RANS

Life developed on Earth, presumably because our planet possessed the appropriate physical properties, supplied suitable raw materials, provided sufficient sparks to initiate the process of evolution, and then was unable to stamp out all the resulting organisms. As yet we have no indisputable evidence of life elsewhere in the solar system, but that could change as our exploratory efforts continue. In this section we will briefly consider the other planets and several other bodies in the solar system, to estimate their potentials for harboring life.

The Moon

The Moon is our nearest celestial neighbor, studied for centuries and first visited by humans a few decades ago. As our model showed, the Moon is considerably smaller than the Earth, with gravity insufficient to hold any significant atmosphere. With no atmosphere or clouds, the Moon's surface prominently displays the results of several billion years of bombardment – a hint of the punishment Earth has absorbed over this same period. The preservation of this impact record is due to the lack of any tectonic motions, current geologic activity, or weathering agents to erase the ubiquitous impact craters. Additionally, the Moon is deficient in volatile compounds such as water, although some observations have indicated that water ice may be present in craters near the Moon's poles. With no atmospheric pressure, liquids vaporize easily, and the smaller molecules launch themselves into space. Life such as ours would be highly improbable without any atmosphere, oceans, or other environmental sources of bodily fluids. And of course, the rocks returned from the Moon by astronauts and automated probes have shown no traces of life, validating what had long been suspected – that the Moon (and any body similar to it) is not likely to support *any* native life forms.

Mercury

The closest planet to the Sun shares many of the same problems the Moon has: essentially no atmosphere, no oceans, no weather, no clouds, etc. Without an atmosphere, one side of Mercury is exposed to the full spectrum of the Sun's rays, raising the surface temperature to far above the boiling point of water. At the same time, temperatures on the night side of Mercury plunge to far below water's freezing point as the surface radiates its energy unchecked to the night sky. Airless bodies such as Mercury and the Moon suffer such extremes in temperature that the task of settling on an appropriate bodily fluid for life would be nearly impossible. This problem is further exacerbated by Mercury's relatively slow rotation rate, which causes a full day/night cycle on Mercury to last 176 Earth days. Flyby missions to Mercury in the early 1970s provided close-up views of a barren, heavily cratered surface, which came as a surprise to no one. All in all, Mercury is not a good candidate for life.

Venus

There was a time when Venus was considered to be Earth's sister planet, being similar in size and distance from the Sun. The fact that Venus has an atmosphere made the thought of life on the planet even more tantalizing. The layer of dense clouds that surrounds the planet prevents our direct observation of any surface features but did spark the imaginations of many astronomers who envisioned a steamy, tropical swamp filled with a variety of plant life. Venus' proximity to the Sun and the discovery that the atmosphere was composed largely of carbon dioxide – one of the raw materials of photosynthesis – contributed further to these speculations.

In fact, the truth is somewhat different. The surface of Venus is *extremely* hot, with temperatures high enough to melt lead. This fact is attributed to a ferocious greenhouse effect caused by the dense atmosphere (about 90 times as dense as Earth's) composed mostly of carbon dioxide. With surface temperatures that high, the water on Venus has long since vaporized; water molecules reaching the top of the atmosphere have been dissociated by the Sun's ultraviolet rays, leaving hardly any in the current mixture. The clouds on Venus – once thought to be water-based – are apparently formed by droplets of sulfuric acid, one of many highly corrosive compounds baked out of the surface rocks by the intense heat.

We have now mapped most of the surface of Venus, using radar from orbiting spacecraft. Results show a number of impact craters – although far fewer than on Mercury or the Moon – and plenty of evidence of volcanic activity, which likely continues today. Photographs taken by space probes sent to the surface show a rather bleak, rocky landscape with no signs of life. None of these probes have survived very long, and it is doubtful that any native life could do any better.

Mars

Although considerably smaller than the Earth, Mars has enough similarities to our planet that a belief in Martians has not required too much stretching of most people's imaginations. Mars' orbit periodically carries it near the Earth, allowing for close observations every two years or so. Its thin atmosphere permits us to monitor its surface, and its rotation period of slightly over a day brings most of the planet into the view of our telescopes on a regular basis. The tilt of its rotational axis is only a degree or two different from that of Earth, indicating that similar seasonal effects might occur on the two planets.

By the beginning of the 20th century, seasonal changes had indeed been observed on Mars. Dark regions on the surface appeared to change size and shape, and polar ice caps could be seen shrinking during the Martian summer – clear evidence (some thought) of the seasonal cycle of plant growth. Additionally, some observers had seen what they called 'canals' – a network of straight line features on the surface, obviously devised by intelligent beings for irrigation and/or transportation purposes. Maps were drawn, books were written, and Martians became very promising candidates for the first alien race to make contact with Earth. Orson Welles helped by broadcasting an adaptation of H. G. Wells' *The War of the Worlds* over the radio in 1938, causing many listeners to believe we were being invaded by Martians.

But it was not to be. As a result of numerous observations by orbiters, landers, flyby missions, and Earth-based telescopes, Mars is now known to be much colder and drier than would be comfortable for us (or Martians anything like us). As in the case of Venus, the primary constituent of the Martian atmosphere is carbon dioxide, but this time the lower atmospheric pressure – less than 1% of Earth's – prevents the development of any significant greenhouse effect to warm the planet. Table 5.1 compares the atmospheres of Venus, Earth, and Mars.

The atmospheric pressure on Mars is too low to hold liquid water on the surface. Even at the low temperatures found there – no warmer than a nice spring day in Minnesota – liquid water should quickly boil away. There is water on Mars, but it is apparently locked away, in the polar ice caps and deep in the soil, perhaps as permafrost. Without access to a supply of liquid water at the surface, life on Mars becomes far less probable.

Table 5.1: Atmospheres of the Terrestrial Planets (% by Volume) (Cox 2000)

	Venus	Earth	Mars
Surface Pressure	90 atm	1 atm	<0.01 atm
CO_2	96.5%	0.033%	95.3%
N_2	3.5%	78.1%	2.7%
Ar	0.007%	0.93%	1.6%
O_2	-	20.9%	0.13%

Mars today does not seem to be the type of planet that should be hopping with life, and it is not. The two Viking missions in 1976, the Pathfinder mission in 1997, and the Spirit and Opportunity rovers in 2004 achieved successful landings on Mars, resulting in a host of close-up images of the surface. None of these show *anything* resembling a living being – no trees, moss, cattle or mosquitoes showed up in the pictures – only a surface strewn with rocks and dust.

Of course microscopic life forms could be present despite the absence of obvious large creatures. The Viking landers carried several experiments designed to look for evidence of current life in the Martian soil. The results of these experiments, while initially appearing to indicate some biological activity, have now been attributed to the weird chemistry of the Martian soil, caused partly by its exposure to ultraviolet rays from the Sun. Although unable to completely rule out the possibility of life presently existing on Mars, the Viking missions have certainly not advanced its cause very far. Martian life may have existed in the past and then died out, but the Viking experiments were not designed to investigate that possibility. Is there any reason to suspect that this scenario might have occurred?

Since 1965 when the first flyby mission to Mars took place, the planet has been studied by a variety of orbiters and passing spacecraft. Visible in the resulting images were impact craters, volcanoes, dust storms, polar caps, and a huge canyon; lacking were the canals, which apparently existed only in the minds of Earth-bound observers. However, there *were* several puzzling features that resemble riverbeds on the Earth, along with other shapes that seem to have been caused by some liquid flowing over the surface in great quantities. The most likely candidate for this liquid is water, based on our previous discussions of abundance, simplicity, and stability. But how could Mars have had plenty of water in the past when it is so dry now?

Like Earth, Mars probably had a warm formation and was geologically active, with volcanic eruptions adding gases to the atmosphere. And, as on Earth, gas molecules launched themselves into space, a process that was more effective on Mars due to its lower surface gravity. Despite these losses, the volcanic outgassing was able to maintain a significant atmosphere, one dense enough to support liquid water on the surface of Mars. It was during this period that the flow patterns and riverbed features were formed.

But Mars was unable to preserve its geologic activity. The smaller planet cooled more rapidly than the Earth has, such that today the Martian volcanoes are inactive and outgassing has ceased. In addition, winter temperatures at the Martian poles are so low that carbon dioxide freezes there, forming polar caps of dry ice (and water ice) and further reducing the supply of molecules available to the atmosphere. With no way to replenish its lost atmospheric gases, Mars will be hard pressed to ever repeat its earlier wet era.

Can we possibly verify that water was indeed the liquid that carved the riverbeds on Mars? One of the principal goals of the missions that landed the Spirit and Opportunity rovers on Mars in 2004 was to search for direct evidence that water once existed on the surface by analyzing the rock formations found there. This search has been largely successful, with a number of different Martian rocks found that appear similar to minerals on Earth that either contain water or require water for their formation. Additionally, layered structures have been found that may well be the result of sedimentary deposits, which would also require liquid water. The exact timing and duration of liquid water on Mars is not yet determined, but its previous existence has become fairly well established (Bell 2005).

Could life have formed during the time when water flowed on Mars, and if so, how far could it have evolved? Evolution of simple life on Mars *may* have occurred, but conditions there probably were considerably less favorable than on Earth. Given the length of time required to produce advanced life forms on Earth (about 4 billion years), it is doubtful whether Martian life could have fared as well. Where should we search for evidence of such life? Regions on Mars where water was most likely to exist would be good places to start. Strangely enough, the *Earth* may provide evidence for ancient life on Mars!

Meteorites have landed on the Earth throughout its history. The sources of most of these rocks from space are the asteroids and comets that populate the solar system, occasionally colliding with Earth. However, a few of these rocks are believed to have originated on Mars (based on the composition of gases trapped in one of them, which matches the Martian atmosphere sampled by Viking). One of these meteorites, called ALH84001, which was found in Antarctica in 1984, has been suspected of actually containing evidence of ancient Martian life forms.

In 1996, scientists proposed that ALH84001 – formed on Mars 4.5 billion years ago, blasted free by an impact around 16 million years ago, and collected from orbit by Earth about 13,000 years ago – contains complex organic molecules, carbonate globules, and fossilized tube structures that are evidence that microscopic life once existed on Mars. Further investigations have not yet verified this hypothesis, as there are other explanations for these observations that do not require the presence of life.

Mars may have once had conditions appropriate for life, but it does not appear to be so hospitable now. Whether life could have evolved there in the past is not clear; but *any* discovery of conclusive evidence of Martian life – living or dead – would be a boost to those who expect to find life elsewhere in the Galaxy. Life on only one planet may be unique; life on *two* different planets could signal a general trend – that life will evolve wherever it gets the slightest chance.

Jupiter

Jupiter is the prototypical Jovian planet, quite different from any of the terrestrial planets previously discussed. The principal difference is composition: while terrestrial planets are formed mostly from iron and rocks, Jupiter is largely hydrogen and helium. In fact, Jupiter's composition is more related to the Sun's than to the Earth's. Because hydrogen and helium make up the vast majority of the atoms in the universe, the Jovian planets made from these materials can be much larger than the terrestrial planets, which have relatively few of these lighter atoms.

Hydrogen and helium are normally gaseous, liquefying only when subjected to extremely high pressures, such as those found in the interior of Jupiter. For this reason, Jupiter and the other Jovian planets are considered to be gas giants. Unlike the terrestrial planets, Jupiter has no solid surface. The 'surface' we observe is actually the tops of the clouds in the Jovian atmosphere; below these clouds lies an ocean of liquid hydrogen and helium that extends to a small, rocky core at the center of the planet. Some molecules of possible interest to terrestrial life can be found floating in the Jovian atmosphere, but there are distinct problems in getting these components assembled into living organisms.

First is the lack of an appropriate liquid medium in which to carry out chemical reactions. In the gas phase, molecules are farther apart and less likely to encounter each other, making reactions proceed more slowly. Second, the Jovian atmosphere is quite turbulent, due to Jupiter's rapid rotation rate. Molecules there are easily carried to different altitudes, where they are subject to a variety of pressures and temperatures. This lack of stability in the molecular environment would make replication of processes and molecules a fairly random proposition. Third, organisms that *might* evolve on Jupiter would have to be able to survive in a wide range of constantly changing environments as they follow the circulation of the atmosphere. With no fixed surface features to cling to, Jovian organisms would be at the mercy of the shifting weather patterns. Lack of environmental stability is likely to destroy life as fast as it might be born.

Thus, although it has plenty of the atoms considered essential to life, Jupiter lacks certain environmental features that would allow life to evolve and succeed. We do not expect to find any significant life forms on Jupiter.

Saturn

Saturn is a smaller version of Jupiter, another gas giant composed mostly of hydrogen and helium. As such, it presents potential life forms with all the same problems that Jupiter does, and its greater distance from the Sun means there will be that much less solar energy available to fuel a living system. Saturn's famous rings contribute nothing to the problem of forming life on the planet. They are actually small particles of ice and rock orbiting about Saturn, but they do not improve the environment for living beings. With no solid surface, no oceans of liquid water, and an atmosphere of dense clouds, Saturn is an unlikely site for life.

Uranus and Neptune

Uranus and Neptune are diminutive gas giants, again with all the attendant problems for life. These two have compositions slightly different from those of Jupiter and Saturn, with a higher fraction of methane, ammonia, and water ices in their interiors. This is not much help for life as these compounds form a slushy mantle beneath a deep global ocean of liquid hydrogen. Their atmospheres are primarily hydrogen and helium with a small fraction of methane, and they show no evidence of biological activity. As a result, we have little hope for the existence of life on these planets that lie at 20 to 30 times the Earth's distance from the Sun.

Pluto

The dwarf planet Pluto is tiny and cold, being so far from the Sun. The Sun's light is so feeble there that Pluto's methane atmosphere is normally frozen solid. There are no compelling reasons why life should exist on such a distant, icy world.

Europa

Jupiter has a number of moons, the four largest being known as the Galilean satellites. Of these, the most intriguing site for life is Europa. Somewhat smaller than the Moon, Europa appears to be covered by a smooth, icy crust that is crisscrossed by a network of strange dark lines. Beneath the icy crust lies a mantle of liquid water, a most appealing discovery to those searching for life. It is supposed that the tidal forces of Jupiter act continually on Europa's rigid crust, alternately fracturing the ice and then pushing it back together. When fractures occur, warmer water is drawn up into the cracks and frozen, forming the dark lines we observe. Whether life could actually evolve in such an environment, close to Jupiter with its hazardous radiation fields, is not yet known, but a large supply of liquid water with fluctuations driven by Jupiter's tidal forces provides an interesting potential for life. A future space mission may send a probe to land on Europa, melt down through the icy crust (perhaps a few kilometers thick), and sample the mantle ocean for life. But while primitive life could possibly exist there, intelligent life is considered highly improbable.

Titan

The only solar system moon to hold a significant atmosphere is Titan, the largest moon of Saturn. Titan is close in size to Mercury – which has *no* atmosphere – but Titan's position 25 times farther from the Sun and its resulting lower temperature considerably reduce the speeds of gas molecules, allowing Titan to retain a dense atmosphere of nitrogen and methane. As part of the Cassini mission to Saturn, the Huygens space probe spent over two hours floating down through this atmosphere and successfully landed on Titan in 2005.

Although Titan's surface is obscured by a smog of organic molecules (generated by sunlight-powered dissociation of methane in the upper atmosphere), when it reached sufficiently low altitudes, Huygens was able to image features that included rivers, streams, and deltas. Unlike on Mars, however, these cannot be attributed to liquid water, which is unlikely to exist at the surface temperatures found on Titan (-290 F = -179 C). Titan is cold enough that the methane there plays a role similar to that of water on Earth; methane clouds form, which produce methane rain, resulting in pools of liquid methane on the surface. This would seem to verify that life on Titan – if it exists – would have to be based on some fluid other than water, making it quite different from any terrestrial organisms. And at the extremely low temperatures found there, chemical reactions – and hence evolution – would have proceeded at a much more leisurely pace than on Earth. We are not apt to be getting any return visits from citizens of Titan.

Asteroids

While there are only a few planets and several dozen moons in the solar system, asteroids (or minor planets) number in the thousands and thus deserve some attention. Asteroids are considered to be rocky planetesimals – bits of rock and metal that failed to be assimilated into a terrestrial planet. They are small, from about one third the diameter of the Moon down to pebble size. (The largest asteroid, Ceres, is now classed as a dwarf planet.) Their low gravities prevent them from holding atmospheres, making them poor sites for life to evolve. Most of them orbit in the **asteroid belt**, between the orbits of Mars and Jupiter; but some have paths that bring them within Earth's orbit, even very close to our planet on some occasions.

The real importance of asteroids in the debate over extraterrestrial life is the effect their *impacts* have had on the evolution and continued existence of planetary life (including ours). There is very good evidence that an asteroid impact 65 million years ago put an end to the age of the dinosaurs on Earth. The giant reptiles were not all hit by rocks falling from the sky; more likely, the impact sent so much debris into the atmosphere that the Earth clouded over, darkening the skies and killing off much of the plant life that was the dinosaurs' ultimate source of food. There is growing evidence that other mass extinctions of terrestrial life may have been caused by similar events, and there is increasing concern that such catastrophes could occur again, with dire consequences for the human species. This aspect of asteroids will be further investigated in a later chapter.

Comets

Comets make up another class of planetesimals, differing from asteroids in their composition, orbits, and appearance. While the asteroids are usually irregularly shaped rocks, orbiting the Sun in reasonably well-behaved paths and not attracting much attention, comets are huge snowballs in elongated orbits, often sporting spectacular tails as they swoop around the Sun during their occasional visits to the inner solar system. The comets are icy planetesimals, comprised of simple molecules of abundant elements – diatomic hydrogen, water, ammonia, methane, etc. – frozen together into icy balls the size of cities. As a comet approaches the Sun, it warms, vaporizing and releasing some of its material, which is pushed outward by the Sun's radiation to form a tail.

Comets play multiple roles in our story. Because of their high composition of volatiles, comets have been suggested as the source of the water for Earth's oceans. Colliding comets would of course contribute their molecules to Earth's supply of water, but whether enough impacts would occur to be able to account for the majority of the water on the planet is unclear. It has also been suggested that comets might have been the *source* of life on Earth, seeding the planet with organisms that formed inside a comet as a result of intense solar radiation acting on the comet's organic molecules during its repeated close passages by the Sun. That an icy comet hurtling through space, its radiation input constantly changing, should provide *better* conditions for initiating life than the Earth itself is an interesting hypothesis, but it has not yet managed to become a convincing theory.

As with the asteroids, the major effect of comets on terrestrial life has probably been their impacts. Being comprised of frozen volatiles, a comet is more likely to *explode* in Earth's atmosphere than form a crater on the surface. Either outcome could have a dramatic, local (or perhaps global) effect on life, possibly affecting the course of evolution. Such an event apparently occurred in the Tunguska region of Siberia in 1908 when a comet (or perhaps a small asteroid) exploded above a remote forest, flattening trees but leaving no crater. Scientists are still studying the Siberian site and speculating on the details of what happened. Had the **Tunguska event** occurred in a more populated region of the world, the results could have been devastating to human civilization.

The Sun

Although the Sun is far too hot to be a home for extraterrestrials, it does provide the gravity to hold the solar system together and the heat to maintain proper temperatures for life on at least one of its planetary bodies. The Sun is a star, viewed from close up, and there are many other stars visible in the night sky. What is the nature of stars? Are other stars just like the Sun? Will they be equally likely to support life on surrounding planets? These questions will be explored in the next chapter as we become better acquainted with stars and the critical role they play in the search for extraterrestrial life.

MAIN IDEAS

- Planets appear to be a natural byproduct of star formation, and it is expected that many – perhaps most – stars are accompanied by planets.

- The planets in our solar system are primarily either terrestrial planets (small, dense planets comprised of rocks and metals) orbiting close to the Sun, or Jovian planets (large, gaseous planets composed mostly of hydrogen, helium, and various ices) located considerably farther away.

- Aside from Earth, none of the other planets or moons in our solar system seem to offer conditions suitable for intelligent life.

- Although we have located large planets around over a hundred other stars, our current detection techniques do not yet allow us to find extrasolar terrestrial planets; this selection effect prevents us from making an accurate observational survey to determine the prevalence of Earth-like planets.

KEYWORDS

asteroid	asteroid belt	brown dwarf
comet	dwarf planet	Europa
extrasolar planets	Jovian planets	Jupiter
Mars	Mercury	Moon
nebula	Neptune	planet
Pluto	Saturn	selection effect
Sun	terrestrial planets	Titan
transit	Uranus	Venus

LAUNCHPADS

1. Other stars are known to have Jovian planets in relatively close orbits about them. Are these stars likely to have terrestrial planets as well? Are such systems likely to have life?

2. Suppose the Earth were completely covered with clouds, as Venus is. Would we have any knowledge of astronomy? Would we be wondering about extraterrestrial life?

3. Could terrestrial life possibly have spread to any of the other bodies in the solar system? How likely is this possibility?

REFERENCES

Basri, Gibor. 2005. A decade of brown dwarfs. *Sky & Telescope* 109(5): 34-40.

Bell, Jim. 2005. In search of Martian seas. *Sky & Telescope* 109(3): 40-47.

Cox, Arthur N. ed. 2000. *Allen's astrophysical quantities.* 4th ed. New York: Springer-Verlag.

Chapter 6
THE DISSIMILAR STARS

*I*n which the nature of stars is described, the HR diagram is introduced, the properties of main sequence stars are examined, stellar evolution and nucleosynthesis are explained, habitable zones around stars are discussed, the Galactic habitable zone is considered, and the complications of binary stars are analyzed.

The night sky is sprinkled with stars; several thousand are visible to the unaided eye on a clear, dark night. While stars appear to us as pinpoints of light – even when seen through a telescope – they are actually immense spheres of hot gas, as is our Sun. Stars share certain characteristics and have similar histories, but they are not all identical or equally supportive of life on the planets around them.

STAR FORMATION

As mentioned earlier in the book, stars form from huge clouds of gas and dust – part of the interstellar medium (the matter not already locked up in stars). The gravitational forces within each cloud concentrate its matter toward the center of the cloud, where it is compressed and heated to make a **protostar** (forming star). The gravitational energy released by the cloud's contraction raises the temperature and pressure in the core of the protostar and slows the contraction, but radiation losses drain some of this energy away. When the core becomes hot enough to ignite nuclear reactions, the nuclear energy released makes the core hot enough to oppose further gravitational contraction and also supplies the energy being radiated away. Once its radius is stabilized in this manner, the protostar has become a normal star; thus, the birth of a star is dependent on the ignition of nuclear reactions.

Like the universe in general, stars are composed primarily of hydrogen (93.9%) and helium (5.9%). The remaining 0.2% of a star's atoms – termed '**metals**' by astronomers – includes all of the other naturally occurring elements. Although only a tiny portion of the star's bulk, this metal fraction is still an important factor in determining whether the star can support life within its planetary system. Metals did not form in any significant amounts during the Big Bang; instead, their creation required the existence of stars and the nuclear reactions that occur within them. Thus, because the elements of life include numerous metals (O, C, N, etc.), nuclear reactions are vital, not only to the stars but also to the life that the stars may support.

Nuclear reactions involve collision and combination of atomic *nuclei*. Normally, the nucleus of an atom is surrounded by an electron cloud, which prevents nuclei from approaching each other very closely. However, deep inside a star the temperature is so high that all the electrons have been stripped from each atom to make a gas of positively charged nuclei and negatively charged electrons. All of these particles zip around in the star at high speeds, colliding with each other frequently; when the right nuclei collide with enough energy, they may combine with each other to form the larger nucleus of a heavier atom. By this process of **nuclear fusion** heavy nuclei are created from lighter ones.

The first major nuclear reaction performed by stars is the conversion of hydrogen to helium. Because hydrogen has an atomic mass of about 1 while helium's mass is about 4, four hydrogen nuclei are required to produce one helium nucleus. In reality, the total mass of four hydrogen nuclei is *slightly greater* than the mass of one helium nucleus; the excess mass is converted into energy – the nuclear energy used by the star to hold itself up and make itself visible to us.

Most of the stars we see in the night sky are generating nuclear energy by converting hydrogen into helium. In later stages of their evolution stars may utilize more advanced reactions to generate energy, converting helium into carbon and oxygen, etc.; the product of one nuclear reaction becomes the fuel for the next as the stars synthesize nuclei of elements up through iron in their quest to harvest energy from the elements. Other elements can be created in reactions that absorb energy, rather than releasing it, such that by a variety of nuclear processes, all of the elements are produced.

There is a key distinction between the *nuclear* reactions described here and *chemical* reactions discussed in previous chapters. Nuclear reactions actually synthesize the nuclei of elements from the nuclei of other elements. Hydrogen nuclei are transformed into helium nuclei, which in turn are transformed into carbon nuclei. These nuclei form the basis of their respective atoms: carbon nuclei are required for carbon atoms, oxygen nuclei for oxygen atoms, and so on. Without the proper nucleus, an atom – or molecules made from that atom – cannot exist; as a result of nuclear reactions, one or more nuclei may cease to exist while one or more other nuclei are created.

Chemical reactions combine existing atoms to form molecules, but they do not produce any *new* atoms. Molecules can be taken apart and their atoms rearranged to form different molecules, but these resulting molecules are still constructed of the same set of atoms as the reacting molecules. For example, when methane (CH_4) combines with oxygen (O_2) to form carbon dioxide (CO_2) and water (H_2O) in the reaction $CH_4 + 2O_2 \rightarrow CO_2 + 2H_2O$, both the reactants and the products contain exactly one carbon atom, four hydrogen atoms, and four oxygen atoms. Chemical reactions occur in our bodies, in the atmosphere, in the oceans, inside the Earth, in the interstellar medium, and even in the atmospheres of some stars, but nuclear fusion reactions are generally confined to the interiors of stars because they require very special conditions.

TYPES OF STARS

Although they may all look like points of light in the night sky, stars are really not all identical. Neither are they all completely different from each other, for stars can be grouped together according to their common characteristics. Stars have many properties, some of which are more easily determined than others. The two most commonly used by astronomers to characterize stars are (1) the rate at which a star radiates energy from its surface and (2) the star's surface temperature. Neither of these properties can be measured directly – we cannot insert a thermometer into a star to take its temperature – but we can deduce these values by measuring other quantities.

We can measure the brightness of a star as seen from Earth, but the value of this measurement will depend on the distance to the star: closer stars tend to appear brighter (other things being equal). However, if we also measure the *distance* to the star, we can combine the brightness and distance measurements to calculate the star's **luminosity** – the amount of energy it radiates each second. (The luminosity of a star is similar to the wattage rating of a light bulb.) Stars with high luminosity radiate more energy, but they might not appear very bright to us if they are relatively far away.

The temperature of a star can be determined in several ways; the simplest method is to observe the star's color. Just as the heating element on an electric stove glows at different colors as it warms up –

typically dull red to red-orange to orange to yellow-orange – stars of different temperatures will radiate different colors of light. The sequence of stellar colors runs from red to orange to yellow to white to blue: cool stars such as Antares and Arcturus appear reddish while hot stars such as Regulus and Rigel appear bluish. (These colors may not be apparent to the casual observer because the eye does not distinguish colors well at low light levels.) By measuring the colors of stars, astronomers are able to rank them in order of temperature.

Another way to measure stellar temperatures involves the spectra of stars. White light is a mixture of all colors; white light that is passed through a prism or grating will be spread out into a rainbow of colors known as a **spectrum**. The spectrum of a light bulb will be a continuous smear of color ranging from red to blue. (Officially, the colors of the rainbow are red, orange, yellow, green, blue, indigo, and violet, but astronomers generally do not use the terms indigo or violet in their work.) The spectrum of a star, produced in the same manner, will also be a rainbow of color, but it will generally not be continuous. As shown in Figure 6.1, stellar spectra typically contain dark lines at certain positions within the rainbow, where individual colors are missing. These lines are due to the absorption of light by particular atoms in the atmosphere of the star. The lines that are observed depend on the composition of the star and also on its temperature because atoms absorb different colors of light at different temperatures.

Figure 6.1: A typical stellar spectrum.

Analysis of stellar spectra is simplified by grouping stars with similar spectra together, and labeling each group with a letter: a **spectral type**. The ordering of the groups is historical, rather than alphabetical; the principal stellar spectral classifications, in order from hot to cool, are O, B, A, F, G, K, and M. Stars in spectral type O are hot and blue while those in spectral type M are cool and red. The designations are further fine-tuned by dividing each spectral type into 10 subclasses. ranging from 0 to 9; the subclasses in type A include A0, A1, A2 ... A8, and A9, followed by F0, F1, F2, etc. The absorption lines in the Sun's spectrum earn it a designation of G2; presumably, other stars in this same spectral class will have the same surface temperature as the Sun.

Thus, a star is assigned a spectral classification based on which absorption lines are present in its spectrum, and this designation indicates the star's temperature, as does the star's color. Temperatures obtained from stellar spectra are in general agreement with those obtained from stellar colors.

Luminosity and temperature are related; other things being equal, a hotter star will be more luminous. However, a star's luminosity also depends on its size; if two stars have the same temperature, the larger one will be more luminous by virtue of its greater surface area. Therefore, a comparison of the luminosities and surface temperatures of stars should also reveal information about their relative sizes.

The standard method for comparing these properties is a graph, called the **Hertzsprung-Russell diagram** after the two astronomers who devised it. The HR diagram is a plot of luminosity versus temperature, with each point on the plot representing a particular star. A typical HR diagram for the stars in our Galaxy is shown in Figure 6.2.

Note that not all combinations of luminosity and temperature exist in stars; the physics of stellar structure clearly favors some relations over others and prevent many more from occurring at all. Note also that there seem to be distinct groups of stars, occupying their own regions of the diagram. Astronomers have named these groups as follows.

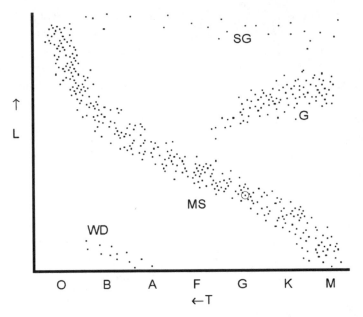

Figure 6.2: The HR diagram.

Most of the stars in the sky fall on a diagonal band running from the lower right corner of the HR diagram to the upper left. This group is called the **main sequence** (MS on the diagram). The next most abundant group lies just above the main sequence, on the right side of the diagram; these are the **giants** (G), which are cooler and/or more luminous than main sequence stars. Across the top of the diagram are found the very luminous stars known as the **supergiants** (SG), while in the lower left corner are the stars that make up the **white dwarfs** (WD). The Sun is a main sequence star of spectral type G2; a small circle marks its position on the diagram.

The names of some of these groups would seem to connote something about the *sizes* of the stars within them. Recall that a star's luminosity depends on its temperature and size; this means that a star's radius can be determined from its position on the HR diagram. A *vertical* line on the HR diagram connects stars with the same *temperature*; because larger stars have higher luminosities, the stars higher up on the vertical line must have larger radii. A supergiant is larger than the giant directly below it, which in turn is larger than the main sequence star of the same spectral type.

A *horizontal* line on the HR diagram connects stars with the same *luminosity*. Comparing two stars on the same horizontal line, the star to the left will be hotter than the one on the right; in order for them to have the same luminosity, the hotter star must be smaller. Thus, a white dwarf is smaller than the main sequence star directly to the right of it, and a main sequence star is smaller than a giant star with the same luminosity. From this analysis it can be seen that the stars increase in radius on the HR diagram from the lower left (the white dwarfs) to the upper right (the cool supergiants).

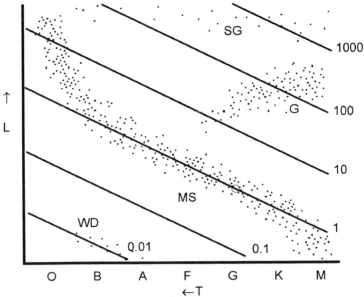

Figure 6.3: The HR diagram showing stellar radii.

Figure 6.3 presents the HR diagram again, with diagonal lines drawn connecting stars of equal radius. The line labeled '1' indicates the stars that are the same size as the Sun, a representative main sequence star. Giants are indeed larger stars, with radii 10 to 100 times that of the Sun, and supergiants are bigger still, from 100 to 1000 solar radii. At the other end of the scale, the white dwarfs are tiny, with radii only 1% of the Sun's. These stars are about the size of the Earth while supergiants exist that are the size of the Earth's *orbit* about the Sun. There is a tremendous range in the sizes, temperatures, and luminosities of stars, and it would be surprising if they were all equally suitable for supporting life.

THE MAIN SEQUENCE STARS

The main sequence is important for several reasons. For starters, it marks the first stage of a star's life after its formation; protostars turn directly into main sequence stars. Because main sequence stars generate nuclear energy by conversion of hydrogen to helium, and because stars have plenty of hydrogen for fuel, the main sequence is the longest, most stable phase of a star's evolution. As the vast majority of stars, including the Sun, are main sequence stars, it makes sense to examine the properties of this group.

Table 6.1 presents the fundamental properties of representative stars from each of the spectral classes. Mass, radius, temperature, and luminosity are all given in terms of solar values; for example, an A0 star has a mass 2.9 times the mass of the Sun, a radius 2.4 times the radius of the Sun, a surface temperature 1.69 times the temperature of the Sun, and a luminosity 47 times the luminosity of the Sun. From these figures it can clearly be seen that mass, radius, surface temperature, and luminosity all decrease down the main sequence from type O to type M.

Table 6.1: Main Sequence Properties (Cox 2000, Allen 1973)

Sp Type	M	R	T	L	Lifetime (yrs)	Sp Type	Abundance
O5	60	12	7.25	400,000	360,000	O	0.00004%
B0	17.5	7.4	5.18	40,000	7.8 million	B	0.1%
A0	2.9	2.4	1.69	47	700 million	A	0.7%
F0	1.6	1.5	1.26	5.7	3.1 billion	F	3.6%
G0	1.05	1.1	1.03	1.3	8.9 billion	G	9.1%
K0	0.79	0.85	0.89	0.45	18 billion	K	14.4%
M0	0.51	0.6	0.66	0.07	54 billion	M	72.1%

Also indicated in Table 6.1 is the main sequence lifetime of each star. Stars do not last forever on the main sequence – only until the hydrogen in their cores is exhausted. From the mass (which gives the star's fuel supply) and the luminosity (which tells how fast the fuel is consumed) we can calculate the duration of a star's main sequence phase. Although the most massive stars have the greatest fuel supply, they burn it at such a furious rate that they do not last very long at all. On the other hand, the least massive stars have such low luminosities that they can essentially last forever – at least several times the current estimates of the age of the universe. The Sun, a G2 star, is estimated to have a total main sequence lifetime of about 10 billion years, about half of which has elapsed so far.

In order to support the development of life on surrounding planets, a star must exist in a relatively stable phase for a minimum length of time. On Earth, life needed about a billion years to get started and another 3.5 billion years to produce a technical civilization. If these times are typical, then we should not expect to find life on planets around stars of types O or B, because these stars do not last more than a few hundred million years.

And if life on Earth evolved more rapidly than normal, we could exclude even more stars from consideration. Of course, if evolution of terrestrial life has been abnormally slow, stars of type F – and possibly even type A – might last long enough to produce civilizations before they run out of hydrogen.

If the requirement of a minimum time for evolution to operate excludes stars of types O, B, and probably A, how many stars are left? Because the more massive stars exist for only a blink of an eye compared to the others, there are not apt to be many of them around. Surveys of the different spectral types show that this is indeed the case, as seen in the final column of Table 6.1. The majority of the main sequence stars are those with lower masses – types K and M. And although far more abundant than stars of types O, B, and A, G stars like the Sun are still in the minority. Obviously, this information will need to be considered as we attempt to determine the number of suitable stars in the Galaxy.

The main sequence is only one group of stars, albeit the majority. What about the other groups? Will any of them provide satisfactory havens for life? The origin of each of these groups can be found in the story of stellar evolution.

THE LIVES OF THE STARS

As has already been related, protostars evolve from gas and dust clouds, and when hot enough, become main sequence stars. The position of each star on the main sequence is determined by its mass, with less massive protostars forming smaller, cooler, less luminous main sequence stars. Every main sequence star generates nuclear energy in its core by the conversion of hydrogen to helium. When this core supply of hydrogen is exhausted, the nuclear energy supply is cut off, making it impossible for the star to support itself with its existing structure. The structural changes that ensue alter the star's surface temperature, radius, and/or luminosity and move the star off the main sequence. The future of a star can be traced by following its resulting evolutionary path on the HR diagram.

Following the main sequence phase, stars follow different paths depending on their masses. For simplicity, we will divide the stars into three mass groups – high (types O and B), medium (types A, F, G, and K), and low (type M) – and describe the evolutionary stages encountered by each group.

At the end of their main sequence phases, medium mass stars such as the Sun undergo a curious transformation. Their newly formed helium cores must be heated before they can ignite as the next nuclear fuel; obligingly, the star's gravity compresses the core, releasing enough gravitational energy to begin raising the temperature. As the core contracts and heats, the star's hydrogen envelope – the portion of the star lying above the core – expands and cools, making the outside of the star appear both bigger and redder. These changes in radius and temperature move the star off the main sequence in the direction of the giant region. Subsequent ignition of the helium core will halt the expansion and stabilize the radius, producing another relatively tranquil – but comparatively brief – life for the star as a **giant**. Giants are generally larger, cooler, and more luminous than their main sequence predecessors.

A star can remain stable as a giant as long as it has a supply of helium in its core to fuse into carbon. When this core helium is exhausted, the star again attempts to solve the problem by compressing its core while expanding its envelope. This time however, the result is a separation of the two: the envelope launches itself into space while the core continues to contract and heat. Ultraviolet rays emitted by the core are absorbed by the gases in the former envelope and then reradiated as visible light. These glowing gases produce a **planetary nebula** in the night sky, such as the Ring Nebula in the constellation Lyra (see Figure 6.4).

The glowing gases in a planetary nebula gradually fade from view as they expand outwards, become less dense, and cease to absorb sufficient radiation from the core. Ultimately they link up with other clouds in the interstellar medium to eventually form new protostars. The core of the former giant

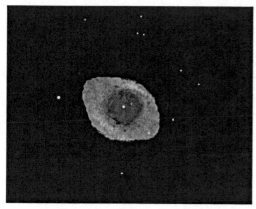

Figure 6.4: The Ring Nebula – a planetary nebula.

continues to contract and heat, but it does not become hot enough to ignite the carbon nuclei that comprise it. Instead, it becomes a very small, hot, dense star known as a **white dwarf**. White dwarfs have no nuclear reactions to generate heat; in fact, their only energy source is the thermal energy stored inside them, which they gradually radiate away over billions of years as they cool. White dwarfs are actually very abundant in the Galaxy, despite being very difficult to see due to their low luminosities.

While the medium mass stars progress through giants, planetary nebulae, and white dwarfs, the high mass stars take a slightly different route. Following their main sequence phase, the high mass stars also undergo a period of core contraction and envelope expansion. But these higher-luminosity main sequence stars become the higher-luminosity **supergiants**, rather than giants. Like the giants, the supergiants will fuse helium into carbon in their cores, but unlike the giants they then go on to more advanced nuclear reactions with much heavier elements as products. The climax comes when the supergiant develops a core of iron nuclei. Because nuclear reactions that use iron as a fuel do not release any energy, the iron core cannot support the star as previous fuels have done. Instead, the core collapses under the tremendous gravitational forces inside the massive star, and this triggers the explosion of the star: a **supernova**.

During a supernova, the contents of the star's envelope are dramatically expelled, eventually merging with the interstellar medium for incorporation into future generations of stars. This process causes a sudden surge in the luminosity of the star, followed by several months of declining light, which should be observable from Earth if intervening dust clouds are not a problem. However, supernovae are relatively rare, as are their predecessors, the high-mass stars; the estimated average rate for the Milky Way is only about two supernovae per century. With the last supernova seen in our Galaxy occurring in 1604, we are probably overdue.

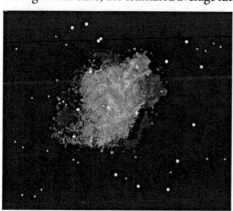

Figure 6.5: The Crab Nebula – a supernova remnant.

Even so, supernovae do leave calling cards for those who may have missed their show. The expanding envelope of gas from the star is lit up by absorption of radiation from the core, forming a **supernova remnant** such as the Crab Nebula in Taurus (see Figure 6.5). Like the planetary nebulae, these faintly glowing clouds will gradually disperse over a few tens of thousands of years, providing a telescopic display for terrestrial – and perhaps extraterrestrial – observers in the meantime. SN1987A, a supernova that occurred in a nearby galaxy in 1987, is still being monitored for the development of a supernova remnant.

The core of the exploded star also has an interesting future. The tremendous gravitational forces involved compress the core so much that protons and electrons are combined to form neutrons. This may result in the creation of a **neutron star**, a dense, spinning sphere of neutrons with the diameter of a large city. In terms of relative size, a neutron star is to the Earth as the Earth is to the Sun. The intense magnetic fields and rapid rotation of these tiny stars cause them to emit pulses of radiation in our direction and reveal themselves as **pulsars**. Neutron stars are quite different from normal stars.

But another fate may await the core of a star that suffers a supernova. If the mass of the core is large enough, the gravitational forces producing the collapse will be strong enough to crush neutrons together, continuing the action until the entire mass of the core is compressed into a point mass. The collapsing core will thus bypass the neutron star phase and become a **black hole**. The gravitational field around a black hole is so intense that even light cannot escape from it (hence the name). Black holes are even stranger than neutron stars, and neither of them is apt to provide much support for extraterrestrial life.

Compared to their larger cousins, the low-mass stars lead rather drab, uneventful lives. At the end of their main sequence lifetimes, these stars will have managed to process almost all of their hydrogen into helium. With no hydrogen envelope to expand, the low-mass star can only contract and heat. But with so little mass, its gravity will be insufficient to ever ignite the helium, and the star will gradually take on the characteristics of a white dwarf. We need not worry about the occurrence of this nonevent, for with main sequence lifetimes of 50 billion years or more, none of these stars have had the chance to evolve this far as yet.

NUCLEOSYNTHESIS

One of the very important functions of stars is the manufacture of the nuclei of heavy elements for subsequent use by future stars, planets, and life forms – a process called **nucleosynthesis**. Stars accomplish this by way of the nuclear reactions that supply the energy the stars need to exist; additionally, some elements are synthesized during the chaos of supernovae, when even energy-absorbing reactions become possible. Despite being only a small minority of the stellar population, the high-mass stars are the most effective at nucleosynthesis. Their high masses and strong gravitational fields allow them to generate the high temperatures necessary for nuclear fusion of elements up through iron, and the dramatic manner in which they end their lives provides a mechanism for producing some of the more exotic elements beyond iron.

It is easy to see why the elements of life – hydrogen, oxygen, carbon, and nitrogen – are among the most abundant in the universe. Hydrogen nuclei, consisting of a single proton, come to us fully assembled from the Big Bang. Carbon is synthesized by the basic core reaction in giants and supergiants, while nitrogen and oxygen are recycled from carbon in the cores of the hotter main sequence stars. Of course, in order for these and other elements to be incorporated into planets and life forms, the stars must return their contents to the interstellar medium for inclusion in the next generation of stars. Had the stars not been able to achieve this goal through supernovae and planetary nebulae, we would not exist.

The initial round of star formation at the beginning of the Milky Way Galaxy would have resulted in stars that were quite deficient in metals. The Big Bang produced essentially no metals, and the stellar factories were not yet operational. Over time, as more stars had a chance to form, synthesize metals, and explode, the interstellar medium gradually increased its metal content. We see evidence of this in the stars that exist today. Those stars that formed long ago, when the metal content of the Galaxy was very low, have spectra that indicate very low metal abundances; these metal-poor stars are called **Population II** stars. Stars that have formed more recently, out of the current interstellar medium with its high metal content are the metal-rich stars of **Population I**. Thus, the metal content of a star is a clue to the star's age; and both metal content and age will be of interest to us in our search.

EVOLUTIONARY HAZARDS

The variety of objects generated by stellar evolution and the events involved with the transmogrification of each star result in a host of hazards for extraterrestrial civilizations. It has already been noted that if civilizations normally require about 4 or 5 billion years to evolve (as ours has), then main sequence stars that lack a stable phase of this duration will be unlikely candidates to support life. Applying the same criterion to the other stages of stellar evolution serves to rule many of them out as well. Giant and supergiant phases are considerably shorter than their main sequence counterparts, making the evolution of civilizations around these stars nearly impossible.

The stellar host must also supply energy to its planets at an appropriate, reasonably constant rate. Neutron stars and black holes are not apt to be satisfactory sources of radiation for planetary life; neutron stars do not produce stable radiation as normal stars do, and black holes do not radiate at all. Once formed, white dwarfs have fairly lengthy lifetimes, but their extremely low luminosities will likely provide insufficient energy to warm surrounding planets. Additionally, the white dwarfs are gradually cooling down, their temperatures changing significantly over time-scales of a few billion years.

Termination of each star's main sequence phase will provide challenges for any life forms existing on surrounding planets. As a star swells up to become a giant or supergiant, the accompanying change in the star's luminosity and/or temperature will wreak havoc on planetary environments. Planets may be roasted – or even worse – engulfed by the expanding star.

A similar, but more violent fate, awaits planets around stars that produce planetary nebulae or supernovae. The hot outrushing gases from either of these events will surely destroy life on nearby planets and strip their atmospheres from them, making the redevelopment of life there highly unlikely. Radiation from supernovae can even have an impact on life around *other* stars, by inducing mutations in their existing species.

The oldest stars in the Galaxy formed long ago, before the abundance of metals had grown to the levels we find today. The nebulae that produced these Population II stars probably had insufficient heavy elements to form many terrestrial planets massive enough to hold the atmospheres needed for life. If this is the case, we should be more likely to find life around the Population I stars, due to their higher metal contents.

In addition to evolutionary hazards, there are several other problems that could also affect developing life on a planet. As the Sun orbits about the center of the Galaxy every 250 million years or so, it may occasionally pass in and out of some of the many clouds of gas and dust that lie in the galactic disk. The effect of these clouds on evolving life or an existing civilization is unclear.

Similarly, in the dance of the galaxies performed by the Milky Way there have been   and continue to be – interactions with neighboring galaxies such as the Magellanic Clouds. These smaller galaxies have passed through the Milky Way's disk as recently as 200 million years ago. There may easily have been other such interactions, for we probably cannot detect all of the nearby galaxies due to obscuration of their light by dust clouds in the galactic disk. Again, the influence these events have had on evolving life here on Earth or elsewhere in the Galaxy is not known, but it has the potential to be quite substantial.

One of the puzzles of modern astronomy has been the nature of the mysterious gamma-ray bursts. Discovered accidentally by Earth-orbiting satellites in the 1970s, these very short bursts of extremely high-energy radiation are detected at the rate of about one per day. The cause of these bursts has been the subject of great debate in the astrophysical community: are they moderately energetic events relatively close by (within the Milky Way), or are they extremely violent eruptions occurring far away across the universe? The hypothesis that is now favored is the latter, with a scenario involving two merging neutron stars emitting a burst of gamma rays of very high energy. The intensity of the radiation produced in these events is so high as to be potentially hazardous for life forms that exist in the same galaxy where the collision takes place. Although apparently quite infrequent – perhaps even *impossible* in a galaxy such as ours – nearby gamma ray bursts might have played a role in determining which species would survive on the Earth.

STELLAR HABITABLE ZONES

Life exists on the Earth, a planet controlled and powered by the Sun; these two bodies have combined to provide an environment in which our biological system thrives. How similar to the Earth-Sun system must another planet and star be in order to also produce and sustain life?

The presence of terrestrial life – and perhaps extraterrestrial life – requires a few minimum conditions, some supplied by the planet, some by the star, and some by a combination of the two. These include the following:

1. An energy source must be available for living beings to tap. (Life requires energy, which must be supplied by the environment to provide living organisms the means of moving, growing, reproducing, etc.)

2. The elements of life – hydrogen, oxygen, carbon, and nitrogen – must be present in sufficient abundance and readily available in the planetary environment. (We have already discussed the fact that terrestrial life is carbon-based and consists primarily of these four elements, and we have made a strong case that extraterrestrial life will have a similar basis.)

3. Water should be abundant on the planet. (Again, water is essential to terrestrial life, and we expect it to play a similar role in alien biologies.)

4. The planetary surface should be maintained within a suitable temperature range in order to keep the water in a liquid state. (Terrestrial life is based primarily on *liquid* water; water vapor on very hot planets or ice on very cold planets will not perform the same function as liquid water.)

5. The planet should have an atmosphere with appropriate characteristics:

   a. It should be dense enough to provide sufficient atmospheric pressure to hold water in its liquid state within the given temperature range (the state of water depends on both temperature and pressure);

   b. It should contain molecules that can serve as a radiation shield, preventing harmful rays from reaching the surface (our atmosphere absorbs ultraviolet rays, gamma rays, and x-rays from space, which might otherwise do considerable damage to us);

   c. It should serve as an insulating blanket, to prevent the night side of the planet from radiating too much heat into space (clouds and greenhouse gases in the atmosphere retard the outward flow of infrared radiation emitted by the Earth and thus keep the surface temperature more uniform over the day/night cycle);

   d. It should contain molecular oxygen (O_2), to allow intelligent species to roam the dry portions of the planetary surface (we walk around at the bottom of an ocean of air, from which we obtain the oxygen we require to breathe).

The principal role of the star is to supply energy to the planet in the form of radiation, providing it at such a rate that the planet's temperature is maintained within certain limits. If the radiation rate is too high, all the liquid water will be vaporized; if it is too low, all the water will freeze solid. (Earth con-

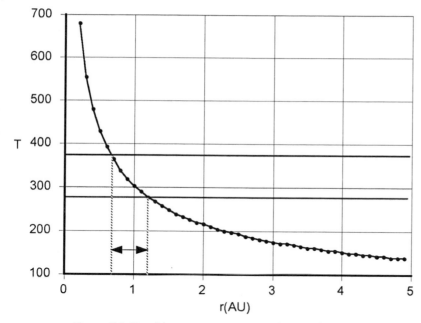

Figure 6.6: Equilibrium temperature vs. distance from star.

tains water in all three states, and it is this balance that is likely to be most beneficial to the existence of life.) The factors that determine the rate of incoming radiation are the star's luminosity, the planet's distance from the star, and the makeup of the planet's surface and atmosphere.

Clearly, a more luminous star radiates energy at a greater rate, resulting in higher temperatures for surrounding planets. Of course, the planet's distance is important too; planets closer to the star will receive a greater flux of radiation, which will tend to make them hotter than planets that are farther out.

The roles played by the planet's surface and atmosphere are not as easy to sort out, due to their dependence on the composition and temperature of each component. We have already mentioned the greenhouse effect (in Chapter 4) and its dependence on the presence of certain molecules in the atmosphere. Another factor is the planet's **albedo**, the fraction of incident radiation that is reflected back into space. A planet with a high albedo will absorb less energy from its star, and this could hold the planet's temperature down. This would be the case for a planet covered in ice, which is highly reflective. A high albedo can also be produced by clouds, but because a cloudy atmosphere may result in a significant greenhouse effect as well, the overall effect on the surface temperature is more difficult to determine in this case.

The effects of luminosity and distance can be seen most easily in a graph. Figure 6.6 is a plot of the equilibrium temperature of a planet as a function of distance from its star, in astronomical units. (One astronomical unit (AU) is the average distance between the Earth and Sun.) This model uses a stellar luminosity equal to the Sun's, a planetary albedo of 0.3 (similar to the Earth's), and a simple greenhouse effect (Carroll and Ostlie 1996) to elevate the temperature. The decrease in temperature with increasing distance from the star is obvious.

The two bold horizontal lines in Figure 6.6 mark the freezing and boiling points of water (measured in degrees Kelvin – the absolute scale commonly used by astronomers) at a pressure of one atmosphere. If we use these extreme temperatures to set limits on life, we can define a range of *distances* from the star within which planetary life may exist, as shown by the two dotted vertical lines. This region around a star where planetary conditions may be suitable for life is called the **habitable zone**. In this case, the habitable zone would range from about 0.66 AU to 1.23 AU, as shown by the arrows. Using different temperatures as limits, a different albedo, and/or a better greenhouse model would, of course, produce different boundaries for the habitable zone.

The effect of the star's *luminosity* on the size and position of the habitable zone can be seen in Figure 6.7. Here the luminosity of the star has been tripled, to three times the Sun's luminosity, moving the curve upward on the graph, away from the previous position (shown in gray). As a result, the habitable zone moves outward, now ranging from 1.14 to 2.13 AU. In addition, the habitable zone also increases its width, from 0.57 AU in Figure

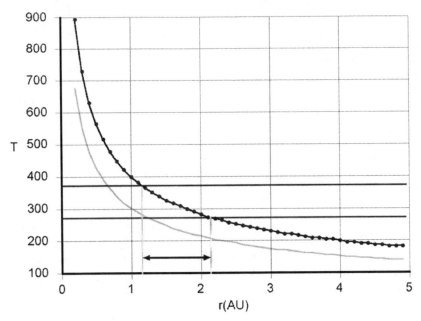

Figure 6.7: Habitable zone for a more luminous star.

6.6 to 0.99 AU in Figure 6.7. Thus, stars with higher luminosities have habitable zones that are broader and farther out; habitable zones of stars with low luminosities form narrow bands close in around the stars.

Figure 6.8 illustrates the variation in the sizes and locations of habitable zones calculated for main sequence stars. Habitable zone *widths* depicted here range from 4.3 AUs for the A0 stars down to 0.06 AU for the M5 stars. (O and B stars have even larger habitable zones, but they are not considered here due to their extremely short lifetimes.) These results are based on the same planetary

parameters (albedo and greenhouse effect) used in the previous two graphs. While changes in these parameters may produce somewhat different values, the general trend is not dependent on the model: the hotter, more luminous main sequence stars have larger, broader habitable zones.

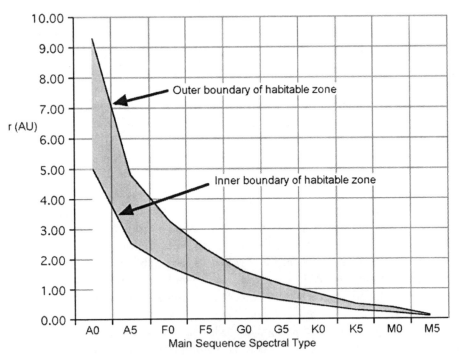

Figure 6.8: Habitable zones for main sequence stars.

The above simple calculations indicate a solar habitable zone about 0.57 AU wide, from 0.66 to 1.23 AU; such a habitable zone would contain the orbits of Venus and Earth. In 1960, similar calculations by Su-Shu Huang (1975) produced a habitable zone ranging from the orbit of Venus (0.723 AU) to that of Mars (1.5 AU) – about 0.8 AU in width. Exercises such as these predict that two or three planets in our solar system should be capable of supporting life. However, with our current understanding of conditions on the Sun's terrestrial planets, it would appear that these calculations are overestimating the size of the Sun's habitable zone. Perhaps there are other factors that should be included in the models.

FINE-TUNING THE HABITABLE ZONE

Development of a technical civilization on our planet required about 4.5 billion years. During this time, conditions on Earth must have remained reasonably favorable for life; otherwise it would have been snuffed out prematurely. Thus, it is not good enough simply to have a planet in the habitable zone – the planet must *remain* in the habitable zone for the several billion years it takes to develop a technical civilization.

There are a number of time-dependent factors that could be considered in order to model the behavior of the habitable zone over time. Examples of these include the following:

1. The luminosities of main sequence stars increase slowly over time, due to the gradual conversion of hydrogen into helium in the core of the star. This increase in luminosity slowly changes the energy input to each planet and moves the habitable zone outward.

2. Formation of oceans on the early Earth initiated the removal of carbon dioxide from the atmosphere (as described in Chapter 4); ultraviolet light from the Sun served to dissociate many of the original atmospheric molecules; and the rise of life on the Earth eventually caused the introduction of free oxygen to the atmosphere. These changes in the atmospheric composition caused a gradual reduction in the greenhouse effect, which in turn, affected the position of the habitable zone.

3. Climatic and seasonal changes on the Earth are linked to the balance of dry land, water, and ice on the surface. This balance, together with the cloud cover, determines the planet's albedo, which affects the absorption of solar energy and thus, the planet's temperature.

4. The duration and severity of seasons depends on several factors that fluctuate with time:

   a. The primary cause of the seasons is the Earth's **obliquity** (or tilt), the angle between the Earth's equator and its orbit. Currently about 23.5 degrees, this angle varies with a period of about 41,000 years. An increased obliquity causes seasons on a planet to be more extreme.

   b. The Earth's orbit is elliptical, rather than circular, making the Earth-Sun distance – and thus the intensity of the Sun's rays as they strike the Earth – a constantly changing value. The degree to which the orbital shape deviates from a circle is measured by the **eccentricity**, a quantity that varies with a period of about 93,000 years.

   c. The intensity of the Sun's rays at a given point on Earth depend primarily on the season and secondarily on the Earth-Sun distance. The changing distance does not *cause* seasons, but it can modify them significantly, depending on the season in which Earth is closest to the Sun. This date is constantly changing due to a motion called **precession**. Precession changes the alignment of the Earth's rotational axis with the long axis of the orbital ellipse, making a complete cycle every 26,000 years. If our summer occurs when we are closest to the Sun, we will have hotter summers and colder winters than we now do.

In the late 1970s, NASA scientist Michael Hart used time-dependent factors to model the evolution of the Earth's atmosphere (Hart 1978). He then used his models to compute the time variation of habitable zones for the Sun and other main sequence stars (Hart 1979). Results of this work illustrate two key points about habitable zones:

First, the increase in the Sun's luminosity over time has served to gradually shift the habitable zone outwards. A planet that might have formed within the *original* habitable zone could find itself too close to the Sun after a few billion years; a planet that appears to be within the *current* habitable zone might have formed outside the original habitable zone, thus reducing the time available for development of life there.

Second, requiring a planet to remain in the habitable zone for several billion years effectively *narrows* the boundaries of the habitable zone, producing a **continuously habitable zone** around the star. The reason for this narrowing effect is illustrated in Figure 6.9.

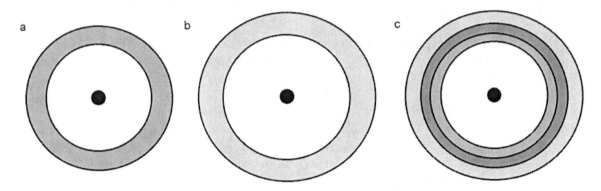

Figure 6.9: The continuously habitable zone.

Figure 6.9a shows the habitable zone of a star (black dot) as a shaded ring around the star; Figure 6.9b shows the habitable zone of the same star several billion years later when the habitable zone is broader and farther out, due to the star's increased luminosity. Figure 6.9c combines the two, with the narrow, dark ring showing the region of overlap that represents the continuously habitable zone around the star. A planet within this dark ring would remain within the habitable zone throughout this period. A planet in the shaded region interior to the dark ring would begin in the habitable zone but be left behind as the habitable zone moved outward; a planet in the shaded region just outside the dark ring would begin outside the habitable zone and wind up in it at the end of the period. The continuously habitable zone is thus narrower than the instantaneous habitable zone; the longer the interval of time involved, the narrower the continuously habitable zone becomes.

Hart's results placed the Sun's continuously habitable zone between 0.96 and 1.01 AU, a width of only 0.05 AU. If the Earth had been formed closer than 0.96 AU, it would have become too warm as the Sun's luminosity increased, possibly evaporating the oceans and destroying life long before it reached an advanced state. If Earth had been formed farther away than 1.01 AU, the early Earth would have been too cool for liquid water to exist; the resulting ice cover would have increased the planet's albedo and prevented the ice from melting, at least for a long time. At best, the formation of oceans and life would probably have been delayed by a few billion years.

If Hart's calculations are accurate, the Earth has walked a very fine line between a runaway greenhouse effect and runaway glaciation during its 4.6 billion-year history, making our presence here seem considerably more remarkable. What about an Earth-like planet around a different star? Insertion of other stars into Hart's atmospheric model produced a similar narrowing of their habitable zones as well. Figure 6.10 shows the results of these efforts compared to the broad habitable zones presented earlier.

In Figure 6.10, the light shading represents the broad habitable zones displayed in Figure 6.8 while the dark band running through the middle of this region marks the continuously habitable zones calculated by Hart. Stars hotter than spectral type F7 were not considered due to their shorter main sequence lifetimes. The cool main sequence stars also pose problems because the widths of their continuously habitable zones dwindle to nothing beyond K0.

Figure 6.10: Hart's continuously habitable zones for main sequence stars (dark shading).

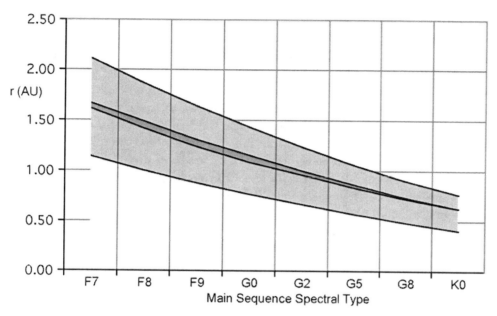

Since Hart's work, other investigators have explored the topic of continuously habitable zones, using different models with additional features. Of particular note, Kasting, Whitmire, and Reynolds (1993) calculated habitable zones by including a feedback mechanism that tends to stabilize a planet's greenhouse effect over long time periods: the **carbonate-silicate cycle**.

The carbonate-silicate cycle links the *rate of change* of atmospheric carbon dioxide to the *abundance* of atmospheric carbon dioxide. This link occurs because atmospheric carbon dioxide is one of the key weathering agents for terrestrial rocks. The chemical details of this process are as follows:

Carbon dioxide (CO_2) dissolves in water (H_2O) to form carbonic acid (H_2CO_3). When rain falls on surface silicate rocks, such as wollastonite ($CaSiO_3$), the atmospheric carbon dioxide and the rainwater combine to weather the rock, resulting in soluble carbonates (HCO_3^-) flowing to or forming in the sea.

$$CaSiO_3 + 2CO_2 + H_2O \rightarrow Ca^{++} + 2HCO_3^- + SiO_2$$

Organic and inorganic reactions transform these soluble carbonates into insoluble carbonates ($CaCO_3$), which form limestone deposits on the seafloor.

$$Ca^{++} + 2HCO_3^- \rightarrow CaCO_3 + CO_2 + H_2O$$

Plate tectonics drives the seafloor under continents at **subduction zones**, heats the limestone ($CaCO_3$), and releases carbon dioxide for outgassing by volcanoes.

$$CaCO_3 + SiO_2 + heat \rightarrow CaSiO_3 + CO_2$$

In this manner, carbon dioxide is constantly cycled from the atmosphere, to the oceans, to the rocks, and back to the atmosphere, at a rate that depends on the atmospheric abundance.

If carbon dioxide levels in the atmosphere are relatively high, they will produce a *strong* greenhouse effect, which will result in *higher* temperatures on the planet. This will increase the evaporation rate from the oceans and generate more precipitation. The greater rainfall rate will result in faster weathering, producing a *more efficient transfer* of carbon dioxide to the oceans and into the rocks. Because subduction and volcanic activity are presumed to transfer carbon dioxide back to the atmosphere at an approximately constant rate, the net effect will be a *decreased* level of atmospheric carbon dioxide.

On the other hand, if carbon dioxide levels in the atmosphere are relatively low, they will produce a *weak* greenhouse effect, which will result in *lower* temperatures on the planet. This will decrease the evaporation rate from the oceans and generate less precipitation. The lower rainfall rate will result in slower weathering, producing a *less efficient transfer* of carbon dioxide to the oceans and into the rocks. Again, because subduction and volcanic activity are presumed to transfer carbon dioxide back to the atmosphere at an approximately constant rate, the net effect this time will be an *increased* level of atmospheric carbon dioxide.

Thus, if the atmospheric carbon dioxide level is high, the carbonate-silicate cycle tends to lower it, and if the level is low, the cycle tends to raise it. In this way the level of atmospheric carbon dioxide is stabilized, and so too is the greenhouse effect, which in turn produces a relatively stable temperature on the planet. The carbonate-silicate cycle adjusts the level of a planet's greenhouse effect in such a way as to maintain a temperature suitable for liquid water; in this way a planet's temperature is made *less sensitive* to its distance from the star, effectively widening the habitable zone.

For the Sun and Earth, Kasting et al. found the inner boundary of the habitable zone at about 0.95 AU – as did Hart. Within this distance the Sun's radiation vaporizes any liquid water, and ultraviolet radiation dissociates the water vapor into hydrogen and oxygen. The hydrogen, being a lightweight element, rises in the atmosphere and ultimately escapes into space. This process effectively removes the water from the planet and makes it very difficult for life to exist.

The carbonate-silicate cycle pushes the outer boundary of Kasting's habitable zone outwards to about 1.37 AU, as shown in Figure 6.11a; beyond this distance carbon dioxide clouds form and increase the planet's albedo until insufficient sunlight is absorbed to keep the planet warm enough for liquid water. Or it could possibly be as far as about 1.67 AU, the farthest point at which the greenhouse effect can prevent the planet's oceans from freezing. In either case, the carbonate-silicate cycle has provided a much wider habitable zone and an increased probability for life.

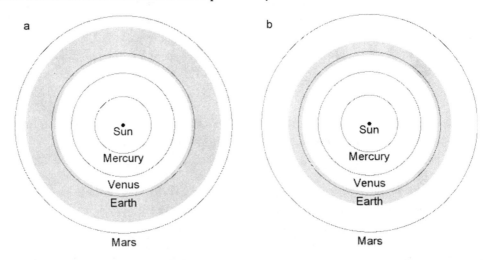

Figure 6.11: The Sun's habitable zone (shaded). (a) Kasting et al. (1993): HZ from 0.95 to 1.37 AU; (b) Kasting (1997): 4.6-billion-year CHZ from 0.95 to 1.15 AU.

Kasting's (1997) continuously habitable zone for the Sun-Earth system (shown in Figure 6.11b) is of course narrower, as it has the same inner boundary as the present habitable zone, but its outer boundary is that of the original habitable zone, 4.6 billion years ago. These values give a continuously habitable zone between 0.95 and 1.15 AU for the 4.6-billion-year interval that the Earth has been in existence – about four times the width calculated by Hart, and it is the Earth's presence in this zone throughout its history that has permitted our evolution on this planet.

Figure 6.12 depicts earlier versions of these habitable zones for the Sun-Earth system. It would seem that the outlook for the Sun-Earth habitable zone is neither as optimistic as Huang first showed (Figure 6.12a) nor as gloomy as Hart later predicted (Figure 6.12b); Kasting's results (Figure 6.11) provide habitable zones and continuously habitable zones of intermediate sizes that have a good chance of harboring Earth-like planets about Sun-like stars. What about other main sequence stars?

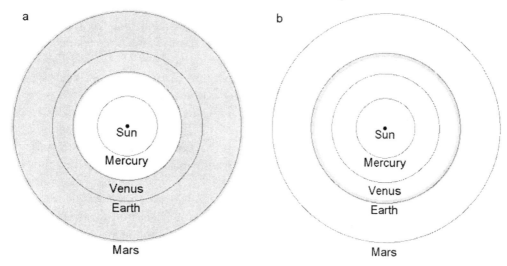

Figure 6.12: Previous models of the Sun's habitable zone (shaded). (a) Huang (1960): HZ from ≈ 0.7 to 1.5 AU; (b) Hart (1979): 4.6-billion-year CHZ from 0.95 to 1.01 AU.

Stars on the upper end of the main sequence have higher luminosities and wider, more distant habitable zones than the Sun; but they also have higher masses than the Sun and thus will evolve more rapidly, with shorter main sequence lifetimes. O stars last only a few million years at best – probably insufficient time for planets to form around them. Lifetimes of B stars – no more than a few hundred million years – are too short to allow life to arise and flourish on surrounding planets. Although life *could* form on the planets of A stars, it will have to progress much more rapidly than it did on Earth in order to produce complex life before the star evolves off the main sequence in no more than about 3 billion years. But the F stars – particularly the cooler ones – certainly stand a reasonable chance of producing technical civilizations, especially if evolution can proceed more efficiently than it did here. Thus, the main sequence stars that are *significantly* hotter than the Sun do not hold much promise for technical civilizations, but those only slightly hotter than the Sun might last long enough to get the job done. The upper main sequence stars are limited by their *shorter lifetimes*, not by their wider, more distant habitable zones.

The lower main sequence stars are a different story. Their lifetimes are considerably longer than the Sun's, providing ample evolutionary time for life. However, the lower luminosities of these stars produce habitable zones that are narrower and closer in to each star, providing considerably more constraints on the formation and orbits of planets that would support life. Additional complications would arise for

115

planets so close to their stars that they become tidally locked, meaning that the planet is unable to rotate freely due to the star's tidal forces. Such planets would wind up with one hemisphere always facing the star and the other forever looking out at the darkness of space.

There are many examples of **tidal locking** in our own solar system. The Moon is tidally locked to the Earth, always keeping the same hemisphere turned toward us; another way to say this is that the Moon's *rotation* period is the same as its *revolution* period. The large satellites of Jupiter and Saturn also show this 1:1 spin-orbit coupling. Although no *planets* in our solar system exhibit this precise effect, Mercury does have a 3:2 spin-orbit coupling, produced by the Sun's tidal forces. The planet completes three rotations in every two orbits about the Sun, resulting in a day/night cycle that is 176 Earth days long with a temperature range of 467 C by day to -183 C by night. (It should not be a surprise that Mercury has been largely ignored thus far in the search for extraterrestrial life.)

Without a reasonably short day/night cycle, the difference in temperatures between the two hemispheres of a tidally locked planet would become quite extreme, making the existence of liquid water – and thus, life – much less probable on such planets. Some have proposed that a dense atmosphere on a tidally locked planet could insulate the dark side and also circulate heat around from the light side, resulting in a less extreme temperature distribution and creating conditions conducive to the formation of life. This may be so, but one could also argue that terrestrial planets that form around low mass stars should be less likely to have masses sufficient to acquire and retain dense atmospheres, due to the decreased availability of planet-building materials in the smaller stellar nebulae that form these stars. This issue is still being debated, but it would appear that while tidal locking does not completely eliminate a planet from consideration in the search for technical civilizations, it probably reduces the chances.

Whether or not a planet becomes tidally locked to its star depends on the initial rotation rate of the planet, the mass of the star, the orbital radius, and time – because it takes time for the tidal forces to slow the rotation rate sufficiently. In general, the closer a planet is to its star, the less time will be required to achieve tidal locking. In the case of the Sun, tidal locking of the Earth at a distance of 1 AU will not be attained in the Sun's lifetime, but for Earth-like planets in the closer habitable zones of less massive stars, tidal locking can be achieved on a time scale much shorter than that required for evolution of life. Such will be the case for stars less than 0.5 solar masses, meaning that planets in the habitable zones of M stars will be tidally locked. What fraction of the main sequence stars are apt to be suitable hosts for life on surrounding planets?

The data in Table 6.1 shows that if we write off the O, B, and A stars, we lose fewer than 1% of the main sequence stars; this will not be crucial to our estimates as our other Drake equation factors are unlikely to be accurate to within 1%. A more important decision concerns the M stars, which account for 72% of the main sequence stars. Eliminating these from contention makes a noticeable dent in the stellar population, and such a move should be considered carefully. If our main goal were simply to find life, no matter how simple, then we should probably include some or all of these stars, but with our particular focus on extraterrestrial *civilizations*, it may be more reasonable to exclude the M stars, given the special conditions that will be necessary to support life of *any* kind on them.

Making the above exclusions leaves us with F, G, and K stars, a total of 27.1% of the main sequence population. With the main sequence stars comprising about 91% of the total stellar population, the fraction of *all* stars that have lifetimes and luminosities sufficiently sun-like to support life on their planets becomes 0.91 x 0.271 ≈ 0.25, or about one star out of four.

RING AROUND THE GALAXY

The concept of a habitable zone around a star is based on the idea that a planet's distance from its star has a major effect on whether the planet will be suitable for life. Some astronomers have proposed that a similar argument can be made regarding the position of a planetary system within the Galaxy – that there is a range of distances from the center of the Milky Way, within which one is most apt to find life. This is the notion of the **Galactic habitable zone**; the reasoning for it is as follows:

Life as we know it is most likely to emerge on terrestrial planets, which are comprised of the heavier elements. But these elements must be manufactured inside massive stars, which ultimately explode as supernovae, enriching the interstellar clouds with metals and making it easier for successive generations of stars to be accompanied by terrestrial planets of reasonable size. In addition, the supernovae trigger additional star formation by compressing the interstellar clouds and inducing their gravitational collapse. The more supernovae that occur, the faster the heavy elements will be made and more likely that terrestrial planets – and hence life – will be able to form.

The highest concentration of heavy elements in the Galaxy is the "thin disk", the region closest to the plane of the disk of our spiral galaxy. But this concentration is not uniform throughout the thin disk, being higher toward the center of the Galaxy and lower toward the outer edges. This is because past supernovae should have been most frequent where the density of stars was greatest – in the central regions of the thin disk. Over time the supernova rate – and the concomitant heavy element abundance – would have increased outward through the plane of the disk, promoting the formation of terrestrial planets in an ever-enlarging zone within the thin disk.

But there is more to the story. While a high supernova rate is good for enhancing metal abundances, it is also dangerous for existing life forms. The interstellar shock waves and intense high-energy radiation produced by supernovae could prove disastrous for complex life forms on planets around nearby stars. (Once life is established on a terrestrial planet, the utility of supernovae diminishes considerably.) This means that the innermost regions of the thin disk, despite an abundance of terrestrial planets, might remain too hazardous for life because of the frequent supernovae; and the outermost region would not have sufficient metal content to produce terrestrial planets. This would leave an intermediate ring around the Galaxy where conditions would be favorable for life: enough terrestrial planets but not too many supernovae. Calculations indicate – not too surprisingly – that the Earth and Sun lie within this ring, called the Galactic habitable zone (Lineweaver, Fenner, and Gibson 2004).

There are additional arguments that can be made concerning the Sun's orbit about the Galaxy and how this orbit may have allowed us to avoid close encounters with many supernovae during the history of the Earth. If such arguments, and the concept of the Galactic habitable zone, are valid, then the Earth and Sun may be even *more* special than we think – and our Drake equation may need another factor.

TWO IS A CROWD

It would seem that the role of stars in the question of extraterrestrial technical civilizations is reasonably well understood by now. We have surveyed a variety of stars, eliminating from consideration those that lead overly violent lives and concentrating on the mundane, slowly evolving, main sequence stars instead. The requirements of habitable zones and lifetimes similar to our Sun's have placed further restrictions on these stars. How comforting a result this is – that our star is a good candidate to support life! But there is another issue that should be addressed here because it affects the nature of a star's planetary system: some stars are really binaries.

A **binary star** is a pair of stars that orbit about each other, held together by the stars' mutual gravitation. The stars may be relatively close together (only a fraction of an AU) and orbiting fairly rapidly, or quite far apart (a few dozen AUs) and moving very slowly. What effect would a binary star have on orbiting planets?

The simple answer is that stable planetary orbits are apt to be much less probable in a binary system. The two stars in the binary will each exert a gravitational force on any planet in the system, and the orbital motion of the stars will continually redirect these forces, making stable orbits difficult to achieve. (Indeed, planetary *formation* might well be less likely in a binary system.) If the two stars are very close together, a planet could orbit successfully at a great distance from them, where their gravitational forces would pull the planet in nearly the same direction. If the stars are quite far apart, a planet might find a stable orbit close around one star where the other star's pull would be relatively weak. But in general, binary stars will not be as conducive to stable planetary orbits as are single stars.

Even if stable orbits exist, there will be other hazards associated with binary stars that may inhibit the development of life on their planets. Habitable zones will be tricky, with two stars of possibly different spectral types supplying energy from constantly varying distances. And continuously habitable zones will be even more complicated as the two stars increase their main sequence luminosities at different rates. Evolutionary events in *either* star, such as progression to a giant phase, will affect life on *all* the planets in the system. And it does not stop there; some stars are really triples, quadruples, or other multiple star systems, which would amplify all of the problems mentioned.

All in all, single stars appear to be far better candidates to support planets with life. What fraction of the stars we see are actually binary or multiple stars? Selection effects prevent us from determining this number exactly, but most estimates range from 40% to 60%. If these estimates are accurate, about half of all stars that would otherwise be considered as good candidates for life may need to be eliminated due to their binary or multiple natures.

Now after all the improbable stars and planets are discarded, how many are left? With so many criteria to meet, both with stars and planets, the answer is not obvious. The solution to this dilemma can be found in Chapter 7, where we will revisit the Drake equation – the tool designed for just this purpose.

MAIN IDEAS

- Stars form from huge clouds of interstellar gas and dust and are comprised primarily of hydrogen and helium, with only a tiny fraction of other elements (metals) present; stars have a wide variety of masses, sizes, temperatures, luminosities, and lifetimes.

- Stellar evolution presents a considerable number of hazards to planetary life around most stars; main sequence stars similar to the Sun may be the only ones with lifetimes and luminosities sufficient for developing intelligent life on surrounding planets.

- Through the process of nucleosynthesis, stars convert hydrogen into helium and heavier elements, including the carbon, nitrogen, and oxygen used in terrestrial life; without a sufficient abundance of metals in planetary systems, terrestrial planets and planetary life may be extremely rare.

- The habitable zone is a region around a star where planetary conditions suitable for life can exist; the continuously habitable zone is an even narrower region in which these conditions can be maintained for the several billion years needed for the evolution of intelligent life.

- Binary star systems may severely limit the probability of stable planetary orbits inside the continuously habitable zones of these stars.

KEYWORDS

albedo	astronomical unit	Big Bang
binary star	black hole	carbonate-silicate cycle
continuously habitable zone (CHZ)	eccentricity	Galactic habitable zone
gamma ray bursts	giant	greenhouse effect
habitable zone	Hertzsprung-Russell diagram	luminosity
main sequence	metals	neutron star
nuclear fusion	nucleosynthesis	obliquity
planetary nebula	Population I/II	precession
protostar	pulsar	spectral type
spectrum	subduction zone	supergiant
supernova	supernova remnant	tidal locking
white dwarf		

LAUNCHPADS

1. Suppose that the luminosities of main sequence stars gradually *decrease* over time, rather than increase. What effect might this have had on life on Earth? On Venus?

2. Our Sun produces mostly visible light, our atmosphere is transparent to visible light, and humans have evolved organs that detect visible light. Are these facts linked to each other or this mere coincidence? Would life on planets around very cool stars develop infrared-sensitive eyes?

3. If supernovae had occurred at a greater rate throughout the history of our Galaxy, would extraterrestrial civilizations be more abundant or less abundant than they are now?

REFERENCES

Allen, C.W. 1973. *Astrophysical quantities.* 3rd ed. London: The Athlone Press.

Carroll, Bradley W., and Dale A. Ostlie. 1996. *An introduction to modern astrophysics.* New York: Addison-Wesley.

Cox, Arthur N. ed. 2000. *Allen's astrophysical quantities.* 4th ed. New York: Springer-Verlag.

Hart, Michael H. 1978. The evolution of the atmosphere of the Earth. *Icarus* 33: 23-39.

Hart, Michael H. 1979. Habitable zones about main sequence stars. *Icarus* 37: 351-57.

Huang, Su-Shu. 1975. Life outside the solar system. In *New frontiers in astronomy*, ed. Owen Gingerich, 104-112. San Francisco: W. H. Freeman. Originally published in *Scientific American* April 1960.

Kasting, James F. 1997. Habitable zones around low mass stars and the search for extraterrestrial life. *Origins of Life and Evolution of the Biosphere.* 27: 291-307.

Kasting, James F., Daniel P. Whitmire, and Ray T. Reynolds. 1993. Habitable zones around main sequence stars. *Icarus* 101: 108-128.

Lineweaver, Charles H., Yeshe Fenner, and Brad K. Gibson. 2004. The galactic habitable zone and the age distribution of complex life in the Milky Way. *Science* 303: 59-62.

Chapter 7
THE RETURN OF THE DRAKE EQUATION

*I*n which the principles of the Drake equation are illustrated by a simple example, strategies for choosing *values of the various factors are explained, sample Drake equations for three individuals are computed, and results of these calculations are interpreted.*

By this time it should be obvious that *every* star will not be a suitable host for life, and *every* planet will not provide an adequate environment for the development of a technical civilization such as ours. The real issue is just how many good sites exist in the Galaxy and how many civilizations have arisen at these locations. Are we alone in the Galaxy, or are there many others?

Chapter 2 described a procedure that can be used to obtain answers to these questions. The technique involves the Drake equation, our version of which is as follows:

$$N = N_* \cdot f_s \cdot N_p \cdot f_e \cdot f_l \cdot f_i \cdot f_c \cdot L/t$$

Observational data provide us with estimates of the various factors, which are then combined to yield a final result. The accuracy of the result depends on the accuracy of each factor used in the equation; a gross overestimate of any particular factor will result in an abnormally large value for N if the other factors are reasonable. Of course, we cannot expect to know *all* the factors precisely, for most of them are estimates at best; but we should be able to pin *some* of them down reasonably well, making our task a bit less difficult.

As mentioned in Chapter 2, we can use the methods of the Drake equation to tackle a number of similar problems. The following example illustrates the thought processes involved in using a similar equation to arrive at a solution.

A DRAKE EXAMPLE

Suppose we need the answer to a complicated – but in this case, rather ridiculous – question: how many female Iowa residents, whose first name begins with E, have a birthday in June and drive a red car with a manual transmission? (Exactly *why* this information would ever be needed is somewhat unclear.) One could of course *guess* an answer without even thinking about the question, and the guess might even be close to the correct answer. But it might just as easily be very far from the truth – perhaps even *obviously* wrong. Application of the principles of the Drake equation should help to prevent such an error.

We begin by composing a Drake-like equation designed to solve this specific problem. The Drake equation itself has factors concerned with stars, planets, and life, none of which have much bearing on the question at hand. We will want an equation with factors that address the criteria given in this particular question.

Our Drake equation begins with a *population* (the number of stars in the Galaxy) and then whittles it down by excluding unsuitable candidates from consideration. For this question we will use a similar approach, beginning with a population (of people, this time) and narrowing the field by applying additional factors. An appropriate population to use might be the population of Iowa, a number found easily in many books and web sites. Let us call this factor P_I.

According to the problem, only *female* residents of Iowa fit the criteria. Therefore the next factor will be the fraction of Iowa residents who are female: f_F. Multiplying P_I by f_F would give the *number* of female residents of Iowa.

Persons meeting the criteria must also have a first name beginning with E. Let the fraction of females with this initial be f_E. Similarly, let the fraction of birthdays in June be f_J.

The final factors are a bit trickier. We can let the fraction of cars that are red be f_R and the fraction of transmissions that are manual be f_M. But because not all females drive cars, we should introduce an additional factor, f_D – the fraction of females who drive a car.

The product of all of these factors should give the answer we seek, which we will call N. The final form of this equation is then as follows:

$$N = P_I \cdot f_F \cdot f_E \cdot f_J \cdot f_D \cdot f_R \cdot f_M$$

All that remains is to determine an appropriate value for each factor, insert these values into the equation, and multiply.

The first number is relatively easy to obtain as the population of Iowa is given in many different sources. The current figure is about 3 million, and that is the value of P_I.

The fraction of females in Iowa is probably about 50%, giving f_F a value of 0.5 (or 1/2).

To determine the appropriate value of f_E, one could list all the different female names, count the number beginning with E, and divide the latter by the total number of names. Of course, listing all possible female names is a lengthy task; a simple estimate might do just as well. As E is one of 26 letters of the alphabet, we could easily use a value of 1/26. This could be modified a bit by arguing that some initials (such as X, Q, Z, Y, etc.) are less common than most, making the remainder *more* probable than 1/26. Perhaps 1/22 or even 1/20 would be a better estimate. Let us take a value of 1/21 for f_E.

The value of the birthday factor f_J is fairly straightforward; there are 30 days in June out of an average 365.25 days in a year. Dividing the two gives f_J = 30/365.25 = 0.082 (or about 1/12).

The fraction of females who drive a car is not particularly easy to pin down, but it should be possible to make a reasonable guess. We should not include those who are too young to drive (estimated at 20% of the population), those who are too old to drive (perhaps 5%), or those who otherwise choose not to drive a car (maybe 10%?). This leaves 65% (or about 2/3) of the female population who drive. (One could argue that some females will drive trucks, rather than cars, but we will assume that fraction to be relatively small and ignore it.)

Red is one of several colors used for cars. We could make an attempt to enumerate all the colors used, but there are many shades of each color, including red, and the list would be long and confusing. A simpler method would be to walk through a filled parking lot and do a survey of the cars found there. For example, 12 red cars out of 100 in a lot would give a fraction of 12%. (One might expect that parking lots in Lincoln, Nebraska would yield a relatively *high* fraction of red cars – another example of a selection effect.)

We can also make a reasonable guess by using a bit of logic. We know from experience that red is a fairly common color; the fraction of red cars is neither as low as 1% nor as high as 50%. Reasonable guesses then might lie in the 5 to 20% range; let us use 10% for f_R.

The final factor is the fraction of manual transmissions. Again, this could be done with a survey of parked cars or a survey of new car dealers, but we will make an educated guess. Manual transmissions are currently in the minority, but this number is nowhere near as low as 1%. A reasonable value might be 10 to 30%; we will set f_M equal to 20%.

With estimates for all of the factors, we may easily calculate the answer by plugging in each value and multiplying them together, as shown.

$$N = P_I \cdot f_F \cdot f_E \cdot f_J \cdot f_D \cdot f_R \cdot f_M$$

$$N = (3 \text{ million}) \cdot (1/2) \cdot (1/21) \cdot (1/12) \cdot (2/3) \cdot (1/10) \cdot (0.2) \approx 79$$

This estimate of 79 female Iowa residents with first initial E who were born in June and drive a red car with a manual transmission is probably not precisely correct, but it is not intended to be. Without more detailed observational data, we could not hope to arrive at the true figure. It is expected, however, that by using a *reasonable* value for each factor, we should have obtained a *reasonable* result, neither outrageously high nor prohibitively low. (An answer in excess of the Iowa population would certainly be deemed to be outrageously high; and the criteria given are probably not so exclusive as to make such individuals nonexistent.)

Some of these factors have very predictable values (P_I, f_P, f_J) while others are more uncertain (f_E, f_D, f_R, f_M). Some of our values may be a bit too high and others may be a bit too low, but ideally these will tend to balance each other and keep our answer in the proper range (whatever that may be). Another person performing this same exercise using reasonable values for the factors should also get a reasonable – but probably different – result.

The same principles apply when using our *real* Drake equation to estimate the number of extraterrestrial civilizations in our Galaxy. We will be making a reasonable estimate of each of the factors in the equation, based on observational data, theoretical speculations, and blind guesses. Unfortunately, for this problem our observational data is considerably harder to obtain; we cannot simply survey parking lots throughout the Galaxy to get the information we need. Values derived from theoretical considerations depend heavily on which theory is being applied (and there are plenty of them). And blind guesses are just that, often based on little more than hunches. As a consequence, the variation in the values of some of the factors will be rather large, producing a correspondingly large spread in the final results. And this is precisely why there is so much uncertainty as to the existence of alien civilizations – we lack the observational data needed to precisely determine many of the factors, select the best of many competing theories, and provide a firmer basis on which to base our wild guesses.

DOING THE DRAKE EQUATION

Although we have not yet covered all of the relevant topics (being only midway through the book), we are now in a position to begin assigning values to the factors in the Drake equation. This exercise should provide the reader with a different view of his or her own thoughts on the question of extraterrestrial civilizations. This view may or may not agree with ideas the reader may have held prior to studying this book, but that is all right. There are many factors to consider in this problem, and it is not always obvious which ones are the most important. Furthermore, given the same set of information, individual readers may still reach widely differing solutions, resulting from their responses to the more uncertain factors in the equation. As with some elections, there will be plenty of room for discussion even after these ballots are counted.

The version of the Drake equation presented in Chapter 2 is reproduced here:

$$N = N_* \cdot f_s \cdot N_p \cdot f_e \cdot f_l \cdot f_i \cdot f_c \cdot L/t$$

As explained earlier, the N to the left of the 'equals' sign is the number we seek – the number of technical civilizations currently existing in our Galaxy. This number can be obtained from the equation after values are assigned to all of the factors. An attempt will be made in this book to indicate an appropriate *range* of values for each factor, at least where such a range can be determined. In each case, the reader is invited to play along, assigning to each factor a value that is consistent with his or her personal views (the Drake equation worksheet in the Appendix may be useful for this project). To illustrate the workings of the Drake equation, we will enlist the aid of three volunteers – Larry, Joe, and Shirley – who will study the information along with us and make their choices for each factor as we progress.

STARTING WITH STARS

The first factor is N_* – the number of stars in the Milky Way Galaxy. This is one of the best-determined factors in the equation, with favored values typically in the range from 100 billion to 400 billion stars. The value most frequently used is probably 200 billion. Larry and Joe both accept this number, but Shirley opts for the high end of the range and takes 400 billion.

SORTING OUT SUNS

The second factor is f_s – the fraction of all stars that are sufficiently Sun-like to support a civilization. We have seen in Chapter 6 that this most likely means main sequence stars within a certain range of spectral types. We can thus eliminate the giants and supergiants (which account for less than 1% of all stars) and the white dwarfs (which make up about 8% of the total). Short lifetimes rule out the main sequence O, B, and probably A stars (less than 1% of the total) while small habitable zones and tidal locking may well take out the M stars (about 66% of all stars). This leaves the F, G, and K main sequence stars, which comprise about 25% of the total (Allen 1973).

Some readers might be content with this value of 0.25 for f_s, but others may wish to include additional factors. We have noted that stars are not uniform in composition: the Population I stars have high metal contents while the metal contents of Population II stars are considerably lower. If we eliminate the Population II stars because their low metal abundances make terrestrial planets and life less probable, we may cut a few percent – probably less than 10% – of the remaining stars from our list. Similarly, concerns about the evolution of life in systems of binary or multiple stars could lead us to discard these candidates as well, reducing our remaining stars by a factor of two. With both of these rather uncertain factors included, the fraction of suitable stars becomes $(0.25) \cdot (0.9) \cdot (1/2) \approx 0.11$.

Given these considerations, we could assume that the value for f_s probably lies somewhere in the range 0.11 to 0.25. Larry opts for an average value of 0.18; Joe, who does not really understand the arguments about metal content or binary stars picks 1/4. Shirley is more conservative, feeling there may even be some *other* factors we have not considered; her choice for f_s is 1/20. (Note that Shirley picked a number that is somewhat outside our selected range. This is not a problem.)

PRODUCING THE PLANETS

The next factor is N_p – the average number of planets per Sun-like star. This is one of many cases where we lack sufficient data to calculate the number. Although we have discovered dozens of extrasolar planetary systems by now, we have done little more than detect the presence of the *largest* planets in each system, and this gives us *no* indication as to the total number of planets there may be orbiting each star.

In fact, the only system for which we have enough data to attempt a reasonably accurate count of the planets is our own, except that this count will depend on just how we choose to define the term 'planet'. Using terminology adopted in 2006, our Sun is considered to have eight planets (from Mercury through Neptune). Whether our solar system is typical in this respect is completely unknown to us at this time.

It is thought that Sun-like stars *should* have planets; most main sequence stars with spectral types F, G, and K spin slowly, indicating that they have transferred most of their angular momentum to their planets. Therefore, the average number of planets is not apt to be less than one. It is also unlikely that planets about a star of this type will number in the dozens; the mass of the nebula from which the system forms, the time available for planets to condense, and the density of stable planetary orbits in a large system would all place limits on the number of planets that could be formed. The range for N_p then might be somewhere between 1 and 20, which leaves our system comfortably in the middle. Larry uses a value of 8. Joe likes planets and thinks 15 is a better number. Shirley is very unsure and decides to guess 5.

EARTH-LIKE ENVIRONMENTS

The next factor is f_e – the fraction of planets that are sufficiently Earth-like to support life. This is a very entertaining factor because although there is considerable observational data concerning the properties of the Earth, it is difficult to determine just how crucial each of these properties has been to the development of life on this planet. This factor will be split into a number of sub-factors, each concerned with a different property; the reader can then assemble a complete factor using any combination of sub-factors desired.

The basic sub-factors include the following:

f_{e1} – the fraction of these planets that are contained in the continuously habitable zones of their Sun-like stars;

f_{e2} – the fraction of these planets that have sufficient mass to hold an atmosphere and slow the loss of internal heat;

f_{e3} – the fraction of these planets whose rotational axis is tilted at an acceptable angle;

f_{e4} – the fraction of these planets with an acceptable rotation rate; and

f_{e5} – the fraction of these planets that have a sizable moon.

The fraction of planets contained in the continuously habitable zones of Sun-like stars depends on the abundances of these stars, the widths of their CHZs, and the spacing of their planets, making a rather complicated problem. For simplicity, we could estimate this value using our solar system as a guide.

As an analogy, consider walking across a darkened room in which eggs have been placed randomly on the floor. What is the probability that you will step on an egg? The bigger your shoes are, the greater the area of floor that your feet cover and the higher the probability of squishing an egg. Also, the more closely the eggs are spaced to each other, the more likely that you will flatten one. Your shoe size relates to the size of the continuously habitable zone in our solar system, and the eggs play the role of the planets. Larger CHZs and/or more closely spaced planets produce a higher probability.

According to Kasting (1997), the boundaries of the Sun's CHZ are at 0.95 and 1.15 AUs, giving a CHZ width of about 0.2 AUs. If the planets about a similar star are spaced at this same interval, we can expect about one planet to be in the CHZ. The Sun's terrestrial planets are located at 0.387, 0.723, 1.000, and 1.524 AUs, leaving gaps of 0.387, 0.336, 0.277, and 0.524 AUs in between, with an average gap of 0.381 AUs, which is greater than the CHZ width. The probability we seek can be approximated by dividing the CHZ width by the average gap size, yielding f_{e1} = 0.2/0.381 ≈ 0.52 or about 1/2.

The mass of a terrestrial planet is important to the development of life because it helps to determine the extent of the planet's atmosphere. We have seen how life on our planet interacts with the atmosphere and how the greenhouse effect was essential to the maintenance of suitable temperatures on Earth during the first billion years or so. We have also seen that the Sun's terrestrial planets have quite distinct atmospheres. Those bodies without atmospheres, such as Mercury and the Moon, are hopeless sites for life.

The key property that determines how much atmosphere a planet can hold is its surface gravity, which depends on the planet's mass and radius. For terrestrial planets, which have similar compositions and densities, mass and radius are linked: larger planets have greater masses. Combining the effects of these variables, we find that surface gravity increases weakly with mass, meaning that more massive planets can retain more extensive atmospheres.

Mass also determines a planet's cooling time. Terrestrial planets that form in a molten state will spend the remainder of their lives cooling as they radiate energy into space. A more massive planet will start with a greater amount of energy (which means more energy must be lost in order to cool to a given temperature), but it will also have a larger radiating surface (which will produce a faster cooling rate). Those planets with a greater surface-to-mass ratio – meaning more surface area for less mass – will thus tend to cool more rapidly. Assuming a similar density for all terrestrial planets, we can approximate this ratio as a surface-to-*volume* ratio, which we can calculate. For a sphere, the surface area is proportional to R^2 ($A = 4\pi R^2$) while the volume is proportional to R^3 ($V = (4/3) \pi R^3$), making the surface-to-volume ratio proportional to R^2/R^3 (or $1/R$). Thus, larger, more massive planets have lower surface-to-volume ratio and tend to dissipate their internal heat more slowly. This more gradual cooling keeps the planet geologically active for a longer period of time, permitting volcanic activity to replenish atmospheric gases that have been lost into space, while also maintaining the benefits of plate tectonics for the planet's life forms.

Venus and the Earth – the two most massive terrestrial planets in the solar system – have retained plenty of atmospheric molecules while Mars and Mercury have been less fortunate. The minimum planetary mass needed should therefore lie somewhere between the masses of Venus and Mars. We will estimate this sub-factor by noting simply that of the four terrestrial planets in the solar system, two seem to have sufficient mass, giving f_{e2} a value of 1/2.

All of the Sun's planets rotate. The angle of inclination of the rotational axis – also known as the obliquity, or the tilt – determines the angle at which the Sun's rays strike each point on the planet and is the principal factor controlling the seasons on Earth. Other Earth-like planets may experience seasonal effects, depending on the values of their tilts. However, if a tilt is so large that the planet's axis is nearly parallel to its orbital plane, the resulting seasons may be too extreme and life may not develop. The factor f_{e3} allows for this possibility by eliminating those planets with unacceptable tilts.

Of the eight planets in our solar system, only Uranus has such an extreme tilt, which would give a value of 7/8 (≈ 0.88) to f_{e3}. However, one could argue that all four of the Sun's terrestrial planets have reasonable tilts, making the sub-factor equal to one. In either case, there will be little effect on the final outcome. We will adopt a value of 0.9 for f_{e3}.

The *rate* of rotation can also be important. A rate too slow may cause the daytime side of the planet to broil while the night side freezes. Most of the planets in the solar system rotate in about 24 hours or less, but two of the eight planets – Mercury and Venus – have spins that are considerably slower. Accordingly, the value for the f_{e4} sub-factor might be in the range 1/2 to 3/4, depending on whether only the four terrestrial planets or all eight planets are used as the basis. Let us assign f_{e4} a value of 0.6.

The fifth sub-factor is perhaps the hardest to judge because the Moon's role in the development of life on the Earth is not obvious. The Moon is the principal cause of Earth's tides, sloshing the oceans on a regular basis and possibly contributing to the development of life along the coastal regions. (The Sun contributes too, but to a lesser extent.) The Moon may also have had an effect on the bombardment of Earth's surface by meteoroids, absorbing some impacts that would otherwise have damaged the Earth. But the most critical function of the Moon may have been to stabilize the Earth's obliquity, preventing its variations from becoming too extreme.

Important or not, these effects would not have been so dramatic had the Moon been much smaller than it is. Most of the moons in the solar system are tiny compared to their parent planets. Based on our solar system, we could estimate the fraction of planets that have a moon of significant relative size to be around 1/8 (the Earth), and that will be the value for our f_{e5} sub-factor, if we choose to use it.

The estimate of f_e then is the product of these various sub-factors. They need not all be used, depending on the whims of the reader; also, others may be introduced as needed if additional properties are deemed critical to the development of life. The value of f_e is then $(1/2) \cdot (1/2) \cdot (0.9) \cdot (0.6) \cdot (1/8) \approx 0.017$ – about 1 in every 60. Larry thinks this sounds too low and uses 1/30. Joe does not want to think of such complexities, deciding simply that if the Earth is the only planet with life out of the eight planets in the solar system, then the factor should be 1/8. Shirley prefers an in-between number, perhaps 1/20.

At this point, we will pause to examine the intermediate results of each of our three volunteers. Multiplying his first four factors together, Larry obtains an estimate of about 9.6 billion Earth-like planets in the Galaxy. Joe has considerably more, with 82.5 billion while Shirley predicts 5 billion. These numbers are actually in reasonable agreement with each other; this is because they are based on fairly good data with individual speculation mixed in. The remaining factors will allow plenty of opportunity for divergence.

THE PRINCIPLE OF MEDIOCRITY

Note that in arriving at values for each of the last three factors (f_s, N_p, f_e), we have relied heavily on our knowledge of the solar system. Lacking sufficient information about *other* planetary systems, we have used the only available example to estimate average values. In essence, we are assuming that our solar system is a typical planetary system. This is an application of the **principle of mediocrity**, which can be stated as follows:

> *There is nothing special or unique about our Sun or the solar system. Our planetary system is typical of those found throughout the Galaxy.*

Note that the principle of mediocrity does *not* say that there is nothing special about the *Earth*. The Earth is *clearly* special, being the only planet in our solar system with a technical civilization. The principle of mediocrity does not even imply that any planetary systems will be *identical* to ours. It simply says that there are going to be other planetary systems with similar – not necessarily identical – groups of planets, some of which will be suitable for life. By using properties of our Sun and solar system to estimate average properties of other systems, we are assuming the principle of mediocrity.

But is it true? Is the principle of mediocrity actually valid for our Galaxy? It probably holds true for the Sun; we see plenty of G2 main sequence stars around the Galaxy and nothing to indicate that the Sun is in any way special (although some astronomers believe the evidence indicates that the Sun has a higher metal content than average). The planets are another story of course; but if we had enough data to know whether this principle is true, then we might not even need to use it. Our solar system could be average in having eight planets, or it could be unique. Eight planets could be abnormally high if the average is actually only one or two planets. But it might be abnormally low if the average is actually 30 or 40. Without the principle of mediocrity, we have very little guidance in estimating the properties of other planetary systems. Unfortunately, that fact does not mean that this principle is valid.

THE LIKELIHOOD OF LIFE

From here on, the values of the factors become extremely uncertain, simply because they all involve life. And although we have *numerous* species of life on Earth, we have only one planet on which life is known to have evolved. The first factor that addresses this issue is f_l – the fraction of Earth-like planets, on which life develops. Given a few billion Earth-like planets orbiting within the habitable zones of Sun-like stars, on what fraction of these planets will chemistry pave the way for the magic biological wand to create what we consider to be life?

The origin of terrestrial life is uncertain, as there were very few eyewitnesses. Raw materials would not have been a problem; experiments by Stanley Miller and Harold Urey in 1952 demonstrated that complex organic molecules (including amino acids) can be made from a mixture of methane, ammonia, hydrogen, and water when an appropriate energy source (such as an electric spark) is applied. The early atmosphere and newly formed oceans could have supplied these conditions.

Life may have arisen in tidal pools, where organic molecules formed by lightning-induced reactions were brought together and concentrated by the drying action of the Sun, eventually resulting in the invention of cellular life. It may have occurred in the depths of the oceans, where organic molecules could be generated by reactions powered by geothermal heat from underwater volcanic vents. We are unaware of any special conditions – beyond those already present on the early Earth – that were needed for the inception of life.

As mentioned in Chapter 4, it has also been proposed that life did *not* originate on the Earth, but rather was delivered here by a comet. This hypothesis sidesteps the problem of generating life on a warm, nutrient-laden, geologically active planetary surface and substitutes the question of how to produce the same thing in a frozen, organic ice ball cruising through interplanetary space. As intriguing as this notion may seem, it does not appear to offer a better solution than concocting life out of the Earth's own home brew.

The bottom line is this: either life evolved on Earth as a natural consequence of the right combination of raw materials, temperature, time, and energy (and would do so again if a similar combination occurred elsewhere), or life was the result of an accidental interaction requiring very specific, rather improbable environmental conditions to be met. In the case of the former, life is a sure thing, perhaps evolving in many locations on a suitable planet; for this case a value of $f_l = 1$ would be appropriate. In the case of the latter, life is an unknown probability. The value of f_l could be 1 in 10, 1 in a thousand, 1 in a million, 1 in a billion, etc., and we are not currently able to tell which is closest to the truth. We are the one chance with life, but we do not know how many other suitable planets must be searched before we find another that has evolved life.

The recommended range for f_l is then from just above 0 to 1; our volunteers are on their own with this one. Larry thinks life will arise occasionally but figures it will be rare, picking $f_l = 0.01$. Joe figures life is always going to happen if the planet is right, and he chooses $f_l = 1$. Shirley thinks Earth life is special somehow and needed an extra spark of some kind to get it started. She has a lot of trouble trying to translate 'special' into a probability, but finally settles on $f_l = 1$ in 10,000. Inclusion of this latest factor leaves Larry with 96 million planets with life, Joe with 82.5 billion, and Shirley with a mere 500,000.

INVENTING INTELLIGENCE

Given an Earth-like planet that has evolved life, what can we say about the development of intelligent species? Does evolution necessarily proceed in the direction of intelligence? Are intelligent species more apt to survive in a changing planetary environment? The answers to these questions will be reflected in the value of f_i – the fraction of life-bearing planets that evolve intelligent beings.

One may argue that intelligent species such as ourselves, being more complex than organisms of lesser intelligence, must therefore represent a pinnacle of the evolutionary process. Hundreds of millions of years of evolution have finally produced us! The time required for Nature to engineer and perfect the human race was so long that it seems to imply that we were the ultimate goal of evolution all along, rather than one of many failed experiments along the way. After all, we do not see another, more intelligent species emerging from the leftover primordial soup, coming to supplant us as the smartest Earthlings on the block. If this is indeed the case, one could argue that any planet in the business of evolving life will keep on working until it finishes the job, producing an intelligent species that will make the planet proud. Such a scenario would lead one to believe that *all* life-bearing planets eventually generate intelligent species and that f_i should be equal to 1.

Alternatively, one could argue that the human species is only one of the millions of species currently existing on Earth. An unknown – and perhaps even greater – number of species that once existed on the planet have by now become extinct. Evolution has created far more *non-intelligent* species that have been (and still are) highly successful at making a living on the Earth, and most of them occupy different branches of the evolutionary tree from ours. Viewed in this way, we are but one of many currently successful species on the planet – perhaps the smartest but neither the most numerous nor the oldest of the bunch. If we indeed owe our existence to the whims of evolution rather than Nature's relentless obsession with intelligence, then thinking beings on other planets may turn out to be a matter of chance. The value of f_i would then be another unknown probability between 0 and 1. We will give our intrepid volunteers some time to ponder this one while we consider the next factor.

TURNING TECHNICAL

The probability that an intelligent species will produce a technical civilization involves a number of interesting points. We have defined a 'technical civilization' as one that has the capability to establish interstellar radio communications. This implies that the intelligent species will need to be able to construct electrical devices suitable for this enterprise. Intelligent beings that live in the sea (such as dolphins) will have a very difficult time *discovering* electricity, let alone working with it. Additionally, intelligent brains housed in the wrong type of body will have real problems getting beyond the theoretical stages of most scientific disciplines. (Flippers are not designed for working with tools.) The body and/or the environment of an intelligent being may conspire to prevent such beings from establishing a technical civilization. (There will be more on this topic in Chapter 10.)

Technical civilizations are almost certainly not automatic for intelligent species. Other than noting that f_c should be less than 1, there is little real guidance for choosing a value; once again our volunteers will have to use their own intelligence to determine this factor, along with f_i. What have they decided?

Larry reasons that intelligence should arise, given enough time, but due to different stellar evolution rates, the time may not be available in every case. As for technology, he figures that a handful of species on Earth could be sufficiently intelligent but most are trapped inside the wrong type of body. Larry sets $f_i = 1/2$ and $f_c = 1/5$.

Joe is not so sure about intelligence. He figures that because there are lots of insects and plants on Earth, that intelligent life will be in the minority elsewhere such that $f_i = 0.01$. He is not quite so pessimistic about technical civilizations, setting f_c equal to 1/10.

Shirley thinks intelligence will arise every time and sets $f_i = 1$. She thinks technology is a reasonable, but not especially probable, outcome and chooses $f_c = 1/20$.

Larry now has 9.6 million technical civilizations, Joe has 82.5 million, and Shirley has only 25,000. Each of these numbers is quite a bit greater than one, implying that the Galaxy can generate a significant number of civilizations around its few hundred billion stars. However, the Drake equation is not finished; there is one more factor remaining to be determined, and it is a very important one.

LASTING A LIFETIME

The final factor in the Drake equation presented in Chapter 2 was f_t – the fraction of time since the development of a planet's *initial* technical civilization that a technical civilization has existed on the planet. In the case of the Earth, this factor would be 1, because our own technical civilization has existed continuously since its inception; however, there may be planets where a technical civilization arose long ago, then died out, and has not been able to regenerate itself, in which case the factor could be very small. To simplify our estimate of this factor, we have chosen to write it as the lifetime of an average technical civilization (L) divided by the average time elapsed since the inception of a technical civilization on each planet (t), giving $f_t = L/t$. While different people will have different ideas about the value of L, we might be able to reach a consensus on the value of t.

The reasoning goes something like this. In a galaxy about 10 billion years old, the maximum star age is 10 billion years. If technical civilizations require a minimum of 4 billion years to establish, then the first one in the Galaxy could have evolved no more than 6 billion years ago. Thus, the maximum value of t is 6 billion years (for the oldest possible civilization), the minimum is about zero (for a civilization such as ours), and if civilizations have arisen at random intervals throughout this 6-billion-year period, the average value of t may be about 3 billion years. Then our factor becomes $f_t = L/(3$ billion years), and we can focus on estimating the value of L.

How long does a civilization last? It is one thing to establish a technical civilization on a planet and quite another to keep it going. A technical civilization requires a supply of natural resources and a reasonably stable environment in order to maintain its status. It must also manage to avoid a variety of perils that may threaten it over the course of its lifetime, and the longer the civilization lasts, the more perils it will face.

Chapter 10 contains a more complete discussion of problems that may place limits on the lifetime of a technical civilization. For now we will note that such things as nuclear war, overpopulation, and asteroid impacts all have the potential to destroy – or seriously damage – a technical civilization, if not the entire human race. Whether alien civilizations will face the same challenges that threaten our civilization is not clear. The aliens may have eliminated many of the problems that plague us but may also have replaced them with even more deadly perils that we have not yet discovered.

Once again, we have little to guide us. We know of only one technical civilization (ours), and we cannot tell how long it will last. Whether its inception is traced to the discovery of radio waves in the 1890s or the initial experiments with radio astronomy in the 1930s makes little difference; we can say that our civilization has been technical for about a century or so. How long we will *remain* technical is much more uncertain.

In our first technical century there was much concern that we would blow ourselves back to the Stone Age with our shiny new nuclear weapons. (The concern is still there, of course, but it is no longer on the front page every day.) If other civilizations also discover radio waves and nuclear energy within a few decades of each other, they too may have to worry about maintaining their newly won technical status through the first century. How far beyond a century they – or we – will be able to last is a subject for the crystal ball.

There are many different perils acting on different time-scales – hundreds of years, thousands of years, millions of years, etc. Some of them are random events while others are more predictable. Some of them have very slow but cumulative effects while others are more sudden and violent. Some of them might be avoidable while others are quite certain. Reading ahead in Chapter 10 might help, or it may just provide the reader with extra sleepless nights.

Once again, our three volunteers are left to their own devices in choosing a value for L, the lifetime of an average technical civilization. Larry thinks that we will soon run out of the natural resources necessary to keep us technical, and overpopulation pressures will make interstellar communication an unaffordable luxury; he sets L equal to 500 years. Joe believes that an intelligent race should be able to solve most of its day-to-day problems, and a technical civilization will be able to turn to alien civilizations in neighboring planetary systems for advice on some of the trickier issues; he figures that an asteroid impact is the most likely end for us and chooses L = 100,000 years. Shirley reasons that global climatic changes pose the greatest threat to our technical civilization, with the next ice age likely to make us concentrate our energy on survival of our species, rather than radio astronomy; accordingly, she picks L = 10,000 years.

COUNTING THE CIVILIZATIONS

The result of the Drake equation is the product of all of the factors selected; simple multiplication provides an estimate (N) of the number of technical civilizations currently existing in our Galaxy. The values selected for each factor by each of the three volunteers and their resulting values of N are presented in Table 7.1. Each person arrived at a different result for N, which is not surprising, considering the variation in the values of their factors.

Table 7.1: Drake Equation Results

	Larry	Joe	Shirley
N_*	200 billion	200 billion	400 billion
f_s	0.18	0.25	0.05
N_p	8	15	5
f_e	0.033	0.11	0.05
f_l	0.01	1	0.0001
f_i	0.5	0.01	1
f_c	0.2	0.1	0.05
L	500 yrs	100,000 yrs	10,000 yrs
N	1.6	2750	0.083

Larry finds N = 1.6, which means that he expects only about one or two technical civilizations to be present in the Galaxy at a time. As *we* are present and technical, one of these would refer to *us*. If Larry's result is correct, we could easily be alone in the Galaxy. If a few other civilizations should happen to exist at this time, they are apt to be so far away from us that we will be very unlikely to discover them.

Joe determines that N = 2750, implying that about 2750 civilizations currently exist in the Milky Way. There may be some hope of locating one or more of these alien civilizations and either traveling to visit them (see Chapter 8) or setting up radio communications with them (see Chapter 9). However, even with this many civilizations, they will still be very far apart; 2750 civilizations spread uniformly throughout the plane of the visible disk of stars would have an average separation of about 1500 light-years. Distances such as these will be explored in the next chapter.

Shirley's result of N = 0.083 perplexes her. She believes that because *we* exist, the value of N should be at least 1. However, there is a way to interpret her result that is perfectly consistent with our presence here. A value of 0.083 for N simply means that a technical civilization exists in the Galaxy for only 8.3% of the time, with *no* civilizations present for the other 91.7%. (An alternative interpretation is to say that a technical civilization exists in only 8.3% of the galaxies at any given time.) With Shirley's result, technical civilizations arise and then normally die out – or at least become non-technical – hardly ever overlapping each other in time. For us, this would mean that we are quite alone in the Galaxy, with even the *existence* of another technical civilization only a very remote possibility, even though there may be an ample supply of planets with non-technical civilizations. After we are gone, the Galaxy will be devoid of technical civilizations for about 110,000 years (on average) until the next eager beings on a different planet around another star begin to encounter the same questions we have been asking. Unfortunately for them, it is unlikely this book will still be around to provide them with any answers.

With these three different results from Larry, Joe, and Shirley, it should be clear that the Drake equation does not favor *either* side in the debate over extraterrestrial civilizations. Solutions to the equation may range from highly optimistic to extremely pessimistic numbers, depending on the values chosen for each of the different factors. The Drake equation is inherently *neutral* – it is just a tool to help assimilate the many pieces of this puzzle into a logical opinion. As long as it does not omit any vitally important factors, it should perform its role quite adequately.

These first few chapters have been devoted to presenting the information we have that bears on the *existence* of extraterrestrial civilizations, allowing the reader to form an opinion as to whether or not we are alone. In following chapters, we will investigate what the existence of alien civilizations would mean to us, in terms of contact, travel, exchange of information, and other interactions. The possibility that such interactions may have already occurred will also be explored.

MAIN IDEAS

- The Drake equation is designed to provide an estimate of the number of extraterrestrial civilizations currently in our Galaxy; similar equations can be used to provide estimates of other unknown quantities.

- Results obtained with the Drake equation may range from extremely pessimistic to extremely optimistic numbers; they depend entirely on the values inserted for the various factors, most of which are not very well determined.

- If it is true, the Principle of Mediocrity may provide guidance in obtaining values for some of the Drake equation factors as it presumes that our Sun and solar system are typical, rather than special.

KEYWORDS

continuously habitable zone (CHZ)
Population I/II
supergiant

giant
principle of mediocrity
white dwarf

main sequence
spectral type

LAUNCHPADS

1. Suppose the Earth had never had a moon. What effect would this have had on the existence of life on Earth?

2. Is it likely that the principle of mediocrity is wrong, that our solar system is unique because it has *only one* civilization rather than many?

3. Suppose that the result of our Drake equation calculation shows that there are one million civilizations in the Galaxy. On the average, how far away should the nearest one be?

REFERENCES

Allen, C.W. 1973. *Astrophysical quantities.* 3rd ed. London: The Athlone Press.

Kasting, James F. 1997. Habitable zones around low mass stars and the search for extraterrestrial life. *Origins of Life and Evolution of the Biosphere.* 27: 291-307.

Chapter 8
INTERSTELLAR ODYSSEYS

*I*n *which the fundamental problem of space travel is explained, the principles of special relativity are illustrated, the pro's and con's of relativistic travel are examined, the nature of UFOs is debated, and the probability of alien visits to Earth is discussed.*

In Chapter 2 it was noted that one way to determine whether aliens exist would be to travel from star to star, scanning each planet for traces of life. It was further noted that this would be rather difficult and time-consuming, given the vast distances between stars. In this chapter we will investigate the subject of space travel in more detail, to judge the feasibility of such a search. We will also attempt to determine whether aliens – if indeed they do exist – are likely to be making such journeys already, even visiting Earth whenever the urge strikes them.

DREADFUL DISTANCES

The major problem with traveling to the stars is that they are all *very* far away. Alpha Centauri, the *closest* star to the Sun, is about 4.3 light-years away – about 270,000 AUs. If the Sun were the size of a BB, the nearest star would be another BB located about 70 miles away. As most of the other stars in the Galaxy are hundreds or thousands of times farther than that, a trip to an extraterrestrial civilization – wherever it is – will not be a trivial undertaking.

Of course, the duration of the trip will depend on the speed of the space vehicle. For a journey at constant velocity, the distance traveled is equal to the product of the velocity and the time; therefore, the time of a trip is equal to the distance divided by the velocity. When measuring distances in light-years and velocities in light speed, the calculation is particularly easy. Light, traveling at 'the speed of light' (c) – about 186,000 miles per second or 300,000 kilometers per second – requires one year to travel a distance of one light-year, 20 years to travel 20 light-years, and so on.

Space vehicles moving at lower speeds will require longer travel times. To cover 20 light-years at a speed of only $0.1c$ (10% of the speed of light) requires $20/0.1 = 200$ years; at $0.01c$ (1% of the speed of light) the same 20-light-year trip will take $20/0.01 = 2000$ years. Currently our space probes do not travel anywhere near this fast. A minimum speed of about 42 km/s ($0.00014c$) is needed to launch a rocket from the solar system on a journey to the stars; at this rate, a trip of 20 light-years requires 140,000 years.

This obviously poses a problem for humans, whose bodies typically last no more than a century. Interstellar journeys that last a significant fraction of a human lifetime are not apt to be very popular. Those that exceed one human lifetime would seem to have some serious logistical issues. Will aliens have the same problem? Could they have life spans that far exceed ours, making interstellar voyages seem no more formidable to them than a trip to Mars appears to us? Without any evidence of the existence of intelligent aliens, we cannot know how long such hypothetical beings might live. But perhaps we can make some reasonable arguments.

Aside from accidents, disease, and other similar problems, humans die of 'old age'. Cells stop reproducing, and parts of the body cease to function (or at least, stop functioning well enough to keep the body going). Apparently our bodies are not preprogrammed to go on living indefinitely. Will alien bodies be any different?

As explained in Chapter 3, we can expect the aliens to be composed of the same elements and use many of the same molecules as we do. Their biochemistry may well be very similar to ours even if their bodies are structured completely differently. If that is the case, they should be subject to the same biochemical limitations on body performance and thus have lifetimes similar to ours (or at least similar to those of other terrestrial organisms), making space travel no more feasible for them than it is for us. Small improvements will not be good enough; due to the vast distances between stars, alien lifetimes would probably have to be hundreds or thousands of times as long as ours in order to make a real difference.

Increased lifetimes may not be the only solution to the dilemma of interstellar space travel. There are some who feel that the laws of physics provide another way to get from here to there without having the journey take forever. All we have to do is travel very, very fast – close to the speed of light – because strange things happen at velocities near the speed of light.

When Isaac Newton developed his laws of motion several centuries ago, scientists were given the power to understand and predict the motions of a wide variety of objects in everyday life. These laws worked equally well for falling objects, rolling objects, and even orbiting objects such as planets, moons, and comets. However, as physicists began to consider objects moving with velocities approaching the speed of light, it was found that Newton's laws no longer yielded satisfactory results. In order to understand the problem – and its solution – we must look at moving objects from different **reference frames** – from different *points of view*.

For example, suppose we are traveling down the freeway in car A at 50 miles per hour (relative to the ground); in other words, in the reference frame of the ground, we are moving at 50 miles per hour. If car B is in the next lane moving in the same direction at 75 miles per hour (relative to the ground), in our reference frame car B would appear to be passing us at 25 mph. And if we find car C in the opposing lane moving *toward* us at 40 miles per hour (relative to the ground), it would appear to be approaching us at 90 miles per hour. These two points of view are illustrated in Figure 8.1.

Figure 8.1a shows the velocities as measured by an observer (X) standing by the roadside. Figure 8.1b shows the relative velocities of the same objects measured by an occupant of car A (who sees car A as stationary and the observer at X moving). The observed speed of any object depends on the **reference frame** from which the observation is made. Converting from one reference frame to another is a simple matter of adding or subtracting speeds – at least it was in Newton's day. Let us now consider measuring the speed of a beam of light.

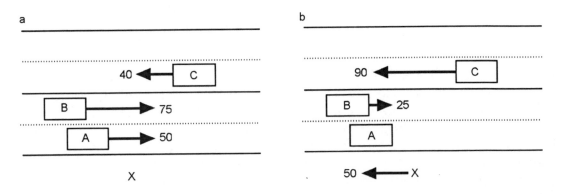

Figure 8.1: Velocities in different reference frames.

Although not an 'object', as such, light does have a speed associated with it, and one of the historical challenges of physics has been to accurately measure the value of this extremely high speed. One might think that if we tried to measure the speed of light, our result would, as above, depend on our reference frame. According to Newtonian mechanics, moving toward the source of the light should tend to *increase* the measured speed while moving away from the source should make the light appear *slower*. But such is not the case: when the experiment is actually performed, we measure the *same* speed for light no matter what the direction or speed of our own motion. This surprising result is contradictory to Newton's laws and suggests that some revision may be necessary to adapt these laws to very high speeds.

SPECIAL RELATIVITY

The revision, developed by Albert Einstein, is called **special relativity**. It has two fundamental postulates:

- Postulate #1: When dealing with motion in a straight line at constant speed, there is no special reference frame – no frame of absolute rest.

All such measured velocities of material objects are relative to the reference frame of the observer, but there is no *unique* reference frame that has zero velocity. In this way all such reference frames are equivalent to each other. Observers in different reference frames, performing experiments involving the motion of objects, would deduce the same fundamental laws of physics.

- Postulate #2: The speed of light is measured to be the same in all reference frames.

Unlike material objects, whose measured velocities depend on the motion of the observer, the speed of light is a constant in all reference frames. No matter what speed or direction the observer is moving with respect to the beam of light, the value measured for the speed of light is always the same. As counterintuitive as this idea may seem, it is in complete accord with experimental results, as noted above.

The postulates of special relativity lead to some interesting effects for objects moving at very high speeds (close to the speed of light). These effects are not normally encountered by humans in everyday life, but they should be considered and understood by potential space travelers.

The first consequence of special relativity is **length contraction**:

In a moving system, objects appear foreshortened in the direction of motion. A rocket whizzing past us at a very high speed will appear shorter to us than if it were at rest in our reference frame.

The second consequence of special relativity is **time dilation**:

Time flows at different rates in moving systems – as speed increases, time slows down. A clock in a rocket whizzing past us at a very high speed will tick more slowly than it would if it were at rest in our reference frame.

To see how these two consequences work, we must introduce some Greek letters – β (beta) and γ (gamma) – define a few terms and do a small amount of math.

Let c = the speed of light, and let v = the speed of another reference frame with respect to us.

Now, let $\beta = v/c$: then as v ranges from 0 to c, β ranges from 0 to 1.

Also, let $\gamma = 1/\sqrt{(1-\beta^2)}$: as β ranges from 0 to 1, γ ranges from 1 to ∞ (infinity).

Then a moving object will have a measured length (L) of

$$L = L_o\sqrt{(1-\beta^2)} = L_o/\gamma \text{ , where } L_o = \text{the \textbf{proper length}.}$$

The proper length is the *maximum* value for L, obtained when the observer is at rest with respect to the measured object.

The duration or time (t) of an event in a moving system will be measured as

$$t = t_o/\sqrt{(1-\beta^2)} = \gamma t_o \text{ , where } t_o = \text{the \textbf{proper time}.}$$

The proper time is the *minimum* value for t, obtained when the observer is at rest with respect to the measured event.

The amount by which the length appears shortened or the time appears lengthened depends on the speed of the moving object (relative to the observer), which affects the values of β and γ as shown in Table 8.1.

At the low speeds common to everyday human life, β is very close to 0, γ is very close to 1, and the effects of special relativity are unnoticeable. *Only at speeds close to the speed of light* do they become significant.

Table 8.1: Relation Between β and γ

β	0	0.01	0.1	0.5	0.8	0.9	0.99	0.999	0.9999	1
γ	1	1.0001	1.005	1.155	1.667	2.294	7.089	22,37	70.71	∞

Relativistic Examples

To observe the effects of special relativity on space explorers, consider a rocket traveling from star A to star B at some speed v. For simplicity, we will assume that Earth and the two stars are at rest with respect to each other.

We on Earth measure the distance between the two stars as the *proper length* because we are at rest with respect to both stars; suppose L_o is measured as 30 light-years. Our other measurements will depend on the speed of the rocket.

If v = 0.8 c, then we on Earth would measure the duration of the trip from A to B as

$$t = L_o/v = 30/0.8 = 37.5 \text{ years.}$$

However, the rocket crew members see things differently. Because they are moving, they measure the distance as

$$L = L_o/\gamma = 30/1.667 = 18 \text{ light-years}$$

and the duration as

$$t_o = L/v = 18/0.8 = 22.5 \text{ years.}$$

This is the *proper time* because the event being measured – the time between their departure from star A and their arrival at star B – occurs in the reference frame of the crew. Thus, *we* see the rocket travel 30 light-years in 37.5 years, but the *crew* think they have gone 18 light-years in 22.5 years.

Both points are equally valid in some sense: while *we* measure the *proper length*, *they* measure the *proper time*. Because time slows down in a moving system, clocks on the rocket – including the biological clocks of the crew – tick at a slower rate and count fewer years elapsed time for the trip. The crew members would actually age more slowly than equivalent humans on Earth. Additionally, the crew's perception of the distance traveled is distorted by relativistic effects, making the trip seem even shorter to them. Thus, the crew members save time in two ways. In this case their reduction in travel time (compared to the Earth's viewpoint) is 40% (calculated from $\Delta t\% = (\gamma - 1)/\gamma = 0.667/1.667 = 40\%$).

The crew can save even more time by traveling faster. Recalculating the results of the same trip for a higher speed of $v = 0.99$ c (giving $\gamma = 7.09$), we find

$$L_o = 30 \text{ light-years and } t = 30.3 \text{ years (measured by us on the Earth), and}$$

$$L = 4.23 \text{ light-years and } t_o = 4.27 \text{ years (measured by the rocket crew).}$$

At this speed, the effects become quite important. The rocket's crew members perceive that they have traveled a *much* shorter distance in a *much* shorter time (86% less) than we would measure. Again, this is a *real* effect, caused by the change in the rate at which moving clocks run and the change in the perception of distance by the moving crew. Thus, the crew can shorten both the interstellar distance and their travel time by moving at a velocity near the speed of light. The closer to the speed of light they can come, the greater will be the effect.

Some may argue that special relativity is 'just a theory', perhaps implying that it is not really valid; but while it is indeed a theory, it appears to be a very good one. Numerous experimental tests of special relativity have been made, and in every case, the results are in agreement with the theory's predictions. For example, there are subatomic particles called muons that can be created by collisions in the laboratory; there they exist for only a tiny fraction of a second before they decay and disappear. These same particles can also be created by collisions between cosmic rays and atmospheric particles, but in this case the muons last about ten times as long as they do in the lab. How can this happen?

The answer is that the muons created in the atmosphere are moving at *relativistic* speeds – over 99% of the speed of light – as viewed from our reference frame. According to the theory, events occurring in the reference frame of the muon (such as the lifetime of the muon) should appear to last significantly longer as measured by observers on the ground, and this is exactly what we see. Thus, the theory of special relativity appears to provide a very good explanation of the behavior of matter at high speeds, and it predicts intriguing benefits for relativistic space travelers, who might be able to travel to the stars within their lifetimes.

Of course there is a catch: the benefits of relativity only work for the rocket crew, not for the folks back home. While the crew members are out cruising from star to star, their friends, relatives, and the financial backers of the journey will all be sitting on the Earth, aging at the normal rate and waiting for results. The crew could return, after what may have seemed to them to be just a few years, only to find that several decades or even centuries have passed on Earth. Would crew members be willing to embark on a journey that – if successful – would effectively transport them forward into the future upon their return, to a time when all of their acquaintances were either considerably older or dead? Would a society be eager to send out such travelers if the society members would all be dead by the time the crew returned?

There is another catch – another effect of special relativity – that must be considered. The *mass* of a moving object increases with its relative velocity. As we accelerate a body closer and closer to the speed of light, it requires more and more energy to gain small amounts of additional speed. The body *seems* to become more massive as its speed increases. The variation of mass with speed is shown in the following equation:

$$m = \gamma m_o \quad \text{where } m_o = \text{the \textbf{rest mass}}$$

Note that mass behaves in the same way time does: as the speed of an object approaches the speed of light, the mass of the object approaches infinity, as does the apparent duration of an event. Thus, material bodies cannot travel *at* the speed of light because they would become *infinitely massive* at that speed. Accelerating a material body all the way to the speed of light would require an infinite amount of energy, something that is usually not available. Material bodies may be accelerated very close to the speed of light – depending on the energy available to do the job – but they can never reach that ultimate speed. The speed of light is the cosmic speed limit on all matter in the universe – a limit that is vigorously enforced.

The cosmic speed limit is not a problem in itself; in theory, the beneficial effects of special relativity – time dilation and length contraction – can be enhanced by pushing closer and closer to the speed of light. In practice, this is a real problem because of the increase in mass with speed.

SPACE VEHICLES

There have been numerous methods of propulsion proposed for space vehicles of various designs, ranging from stellar sails (which use the light from stars to push them through space) to interstellar ramjets (which scoop up interstellar gas and shoot it out behind at high velocities) to spaceships powered by the thrust of nuclear bombs, exploding against a shield at the rear of the craft. A thorough discussion of the technical merits of different propulsion systems will not be attempted here. However, we *will* investigate the effects of special relativity on the device used to propel our current space vehicles – the rocket.

A typical rocket carries fuel and an oxidizing agent, which is used to burn the fuel. The gases that are the products of this combustion reaction are then exhausted out the rear, forcing the rocket forward (due to Newton's third law of motion: *For every action there is an equal and opposite reaction*). When the rocket engine is ignited, the rocket accelerates; when the engine is turned off, the rocket coasts in whatever gravitational field is available. In deep space, far from any stars, the rocket would cruise in a straight line at a constant speed unless its engine were fired to change its velocity.

In its simplest form, a rocket consists of payload and fuel. The payload is the part of the rocket being delivered to the destination, including the crew, the cargo, and any structural elements necessary to move them. The greater the mass of the payload, the more fuel must be expended to accelerate it to cruising speed.

Suppose we assume maximum fuel efficiency by relying on total conversion of matter into energy and ejection of photons out the rear of the rocket to generate the thrust. With this process (which is tricky in practice) there will be no waste – no unused fuel. We can then use the following equation to calculate the total mass (M) of a rocket (payload plus fuel) required to bring a certain payload mass (m) to the desired speed (specified by β):

$$M = \sqrt{[(1+\beta)/(1-\beta)]}\ m$$

For example, accelerating a rocket to a speed of 0.8 c (β = 0.8) would require a total mass M = $\sqrt{[1.8/0.2]}$ m = 3 m, meaning the fuel would have twice the mass of the payload. Acceleration to 0.99 c would require a total mass M = $\sqrt{[1.99/0.01]}$ m $\approx$ 14 m, meaning the rocket must carry 13 times as much fuel as payload. As speeds increase further, the fuel-to-payload ratio increases dramatically, making design and construction of such vehicles a real engineering problem. Unfortunately, the problem has further complications.

In making a journey to another star system, the rocket might first accelerate to get up to cruising speed, then cruise at constant speed to the target star, and finally decelerate as it nears its destination. The return trip will add another sequence of acceleration, cruising, and deceleration, for a total of four accelerations for the round trip. (The *process* of deceleration is the same as that for acceleration except for the direction the rocket engine is pointed.) Thus, a trip to another star and back would require a minimum of four expenditures of fuel.

The fuel required for *future* rocket burns must be accelerated along with the payload, and in essence, becomes part of the payload until it is burned. Thus, the payload for the first acceleration includes the fuel for the second, third, and fourth accelerations, and the payload for the second acceleration includes the fuel for the third and fourth accelerations, and so on. The total mass of a rocket (payload plus all fuel) required to supply the four burns occurring on a roundtrip voyage is given by the following equation:

$$M_4 = [(1+\beta)/(1-\beta)]^2 \, m$$

Results of these calculations for various speeds are shown in Table 8.2; for each speed (shown as β), the table presents the ratio of total rocket mass to payload mass, for one acceleration (M/m), and for the four accelerations required for a round trip (M_4/m).

Table 8.2: Fuel Requirements for Relativistic Rockets

β	M/m	M_4/m	Relativistic Time Savings
0.1	1.11	1.5	0.5%
0.5	1,73	9	13%
0.8	3	81	40%
0.9	4.36	361	56%
0.99	14.1	39,600	86%
0.999	44.7	4 million	96%
0.9999	141	400 million	98.6%

Also shown is the time saved for each trip due to relativistic effects, a comparison between the elapsed trip time experienced by the crew and the trip duration as measured by the folks back home. For a significant savings in travel time, a near-light speed is essential. Unfortunately, the fuel requirements would appear to become prohibitive at such speeds. For example, to travel at 99% of the speed of light – which would cut travel time by 86% – would require a rocket on a roundtrip journey to carry about *40,000 times* its payload mass in fuel. Clearly there is a heavy price to pay for relativistic rocket travel.

Of course the situation is not so bleak if the space vehicle does not have to carry its own fuel. Perhaps there will be interstellar refueling stations dispensing the most popular brands of rocket fuel at competitive prices. Perhaps they will even accept major credit cards; perhaps not.

Or it may be that a spacecraft that does not carry its own fuel *can* be successfully designed and produced. The interstellar ramjets mentioned above would need to be immense and/or have very powerful magnetic fields to collect enough interstellar gas particles to provide their propulsion. Similarly, the light sails, which would use radiation pressure from the stars to propel spacecraft, must have a very large surface area in order to catch enough light. The sizes necessary for these spacecraft present two more problems: increased size means increased payload mass to accelerate; and extremely large spacecraft require huge quantities of raw materials to produce and must be assembled in space. Construction of interstellar spacecraft will be a tremendous undertaking for whatever civilization decides to make the attempt. It is not obvious that every technical civilization will do so.

RELATIVISTIC PITFALLS

Design and construction of interstellar spacecraft are not the only challenges awaiting civilizations that would journey to the stars. There are problems involved in the operation of such craft that are exacerbated by near-light speeds:

There are many types of objects in the Galaxy: huge clouds of gas and dust, stars, brown dwarfs, planets, comets, asteroids, meteoroids, and atomic particles, to name a few. Although the stars are reasonably visible to the eye, the other objects become increasingly difficult to spot – and also more numerous – as their sizes diminish. Meteoroids are generally not detected until they crash into the Earth's atmosphere, when the frictional heat generated produces the flash of light we see as a shooting star or meteor. Similarly, such objects in deep space will not be noticed until they collide with something – such as a space vehicle.

It would be hoped that designers of the spacecraft would have provided it with a skin tough enough to ward off such intruding particles. Of course, every material has its limits, and the spacecraft cannot be expected to carry sufficient armor to deflect really *large* particles (which should also be less abundant). Most likely, the spacecraft will be produced with 'five-mile-per-hour bumpers' designed to repel the most numerous small particles but providing inadequate protection against more violent collisions.

Now consider a relativistic rocket, cruising at near-light speed through the Galaxy. Far from any stars or other large, bright objects, the crew would probably not notice their own motion, just as the observers in car A in Figure 8.1b did not perceive their own velocity. However, just as the observers in car A viewed the (stationary) observer at X as moving past them (with the same speed car A had relative to X in Figure 8.1a), the rocket crew would find that stray particles in space would now appear to be whizzing past them at relativistic speeds due to their own high rate of speed (as shown in Figure 8.2).

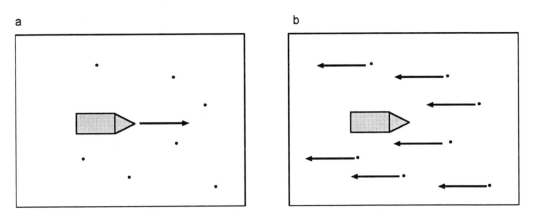

Figure 8.2: Relativistic space debris in the reference frame of (a) the debris and (b) the rocket.

At very high spacecraft speeds, every particle of space debris becomes a relativistic bullet with the potential for doing great damage to the spacecraft and its crew. Cataloging the position and velocity of every particle in the path of a relativistic spacecraft in hopes of predicting their future positions and dodging them all would seem to be an impossible task. Perhaps there will be spacecraft body repair shops along the way.

Another serious problem involves relativistic navigation. At near-light speed, one's view of the surroundings is severely distorted; apparent positions of stars are shifted by the rocket's motion. As the rocket's speed increases, the forward hemisphere (and an increasing fraction of the backward hemisphere) of the sky is compressed into a narrowing cone in front of the rocket. Stars that would normally be seen on either side of the non-relativistic rocket would appear to be shifted around to this front cone as the rocket accelerated. These changes would clearly alter the familiar patterns of the constellations and other star groupings, making identifications and determinations of the rocket's position extremely difficult.

Furthermore, the color and brightness of a star's light can also be changed by relative motion between the star and the rocket, which causes the star's radiation to shift its position with respect to the visible spectrum. We can approximate this portion of the spectrum by laying out the colors of the rainbow in order, as shown in Figure 8.3; note that the regions adjoining the visible spectrum (ultraviolet and infrared) have also been included.

Because of their characteristic temperatures, stars emit most of their radiation in these three spectral regions; Figure 8.3 shows a simplified spectral distribution curve for the Sun, which emits its peak radiation in the middle of the visible spectrum. A hotter star will have a radiation peak to the left of the Sun in Figure 8.3, perhaps in the ultraviolet region as shown, while a cooler star will have its peak to the right, possibly in the infrared region. Note that each star emits radiation over a broad spectral range.

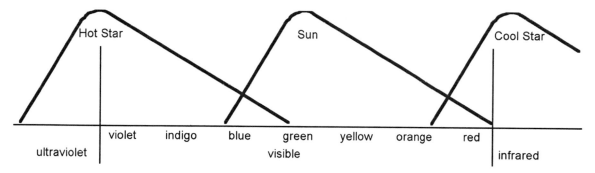

Figure 8.3: Effect of stellar temperature on stellar color.

The brightness of a star depends on the amount of its radiation that is emitted in the visible region of the spectrum (because the eye cannot detect ultraviolet or infrared rays). The color of a star depends on the amount and color of this visible radiation. For example, in Figure 8.3 the visible radiation from the hot star is primarily violet to blue in color, and such stars normally appear bluish; for similar reasons, the cool star appears reddish. The Sun – and a star of the same temperature – radiates a broad range of visible colors that combine to make them appear essentially white.

As noted above, relative motion between the star and the observer changes the star's appearance by shifting its radiation along the spectrum – a phenomenon known as the Doppler shift. If the observer is approaching a star, the star's radiation will be shifted to the left in Figure 8.3, toward the bluer colors; this is known as a blueshift. Motion of the observer away from a star produces a shift to the right – a redshift. The greater the speed of the observer, the greater is the shift of the radiation. At the normally low velocities with which we are familiar, the shift is barely noticeable, but at relativistic speeds, the effect can be dramatic. At a velocity half the speed of light, the shift is equivalent to the entire length of the visible spectrum!

If the shift *increases* the amount of a star's radiation in the visible, the star will look brighter, but if the shift *decreases* a star's visible radiation, then it will look fainter. The color of a Doppler-shifted star will depend on the distribution of its shifted radiation within the visible spectrum, but because the relativistic shifts can be so great, there is no simple statement that will cover all stars.

Furthermore, not all of the 'stars' seen by the crew of the speeding rocket will actually be stars. The relativistic blueshift bestowed upon objects seen in the forward direction is capable of moving sources from the infrared into the visible where they may masquerade as stars; meanwhile, the *real* stars will have been blueshifted into the ultraviolet, where they will be largely invisible to the human eye.

Thus, the objects observed visually by the crew of a relativistic spacecraft will be a confusing collection of old and new sources, with brightness, color, and spectra made less recognizable by the Doppler shift and stellar positions distorted dramatically. Successful navigation at relativistic speeds will probably require a detailed knowledge of *all* the objects along the route *before* the trip is made. Unfortunately, such information will be difficult to obtain before actually making the trip. Without proper preparation, the first relativistic adventurers may very well become lost in space.

Visitors from Space?

It would seem that a civilization's decision to travel to the stars is going to require a major commitment of technology, energy, and time. Even *with* such a commitment, these efforts may still not be sufficient to overcome the limitations of physics and the reality of astronomy. In short, interstellar travel just *may* be impossible.

At the same time, there are thousands of reports every year of strange lights in the sky, unexplained marks on the ground, mysterious abductions, and other occurrences that are taken by many humans to be direct evidence of the presence of extraterrestrial intelligent beings on Earth. If we are, in fact, being visited, then the above discussion of space travel must have some big holes in it. If we are *not* being visited, then there must be better explanations for the various phenomena being reported.

Many individuals are convinced that UFOs are evidence of alien spaceships. A **UFO** is an **Unidentified Flying Object**, meaning an object or light seen in the sky that cannot be identified by the observer. UFOs are relatively common because there are *lots* of things in the sky, and most people do not pay much attention to them. When they *do* notice a bright light in the sky, their experience may not provide them with enough options to build a satisfactory terrestrial or celestial explanation, thus leading them to the default value – extraterrestrial spaceship.

Most astronomers are well aware of the variety of objects that populate the sky, and they are especially sensitive to *lights* in the night sky. The Moon, stars, planets, comets, meteors, aurorae, satellites, airplanes, helicopters, balloons, birds, bats, fireflies, and clouds can all masquerade as UFOs if conditions are right. It would make sense to thoroughly explore these less-exotic explanations before resorting to alien spaceships because identifying UFOs as extraterrestrial spacecraft raises a host of other rather difficult questions.

The presence of extraterrestrials here on Earth requires (1) that an alien technical civilization has evolved on another world, (2) that the alien civilization has developed the means for interstellar space travel, (3) that the aliens choose to spend at least some of their time and resources traveling to other stars, and (4) that they have in fact come to the Earth. Assuming that the first three requirements are true – that there are spacecraft actually traveling among the stars – why would they come to the Earth? How would they even know we exist?

A Terrestrial Homing Beacon

It could be that they are here by accident. Perhaps in their voyages through the Galaxy they just stumbled upon the Earth. This must be regarded as highly improbable, given the immense size of our Galaxy and the enormous numbers of stars it contains. In Chapter 7, we saw that estimates of the number of Earth-like planets in the Galaxy ranged in the *billions*, making odds against a purely accidental discovery of our planet on the order of a billion to one at best. If our planet or star had something to attract the aliens to us, that might make things easier.

As far as we know, there is nothing special about our star (unless it should turn out to be true that its metal content is abnormally high for a G2 main sequence star, as some astronomers suspect). In fact, the principle of mediocrity assumes just that — that the Sun is one of billions of similar stars, none of which should attract any undue attention. For its part, the Earth is a tiny speck of rock, reflecting only a microscopic fraction of the Sun's light; surely, the Earth will not stand out among the billions of planets from which the aliens can choose. However, as strange as it may seem, the Earth just might do that.

Within the solar system, the Earth is the only planet with an atmosphere that contains significant amounts of free oxygen, placed there by the action of living plants over billions of years of evolution. Because the composition of a remote planetary atmosphere can be deduced — at least in part — from the spectrum of light we receive from it, we have *some* information about the components of other planetary atmospheres in our solar system.

We could do the same for the atmospheres of planets around other stars, except for a few problems. We cannot yet see other terrestrial planets directly because the starlight they reflect is too faint for us to detect, and their proximity to their parent stars prevents us from resolving separate images of them. With an ambitious program to back construction of bigger and better telescopes, we may ultimately be able to obtain images and spectra of planets around nearby stars. Results of these observations may yield planets that also have atmospheres with high oxygen contents, making them good candidates for advanced life, and while *we* do not yet have these capabilities, an alien civilization may, leading them to consider the Earth as a promising planet to visit.

Although making it more probable, free oxygen in a planetary atmosphere does not *guarantee* the presence of an intelligent civilization. But there *is* something that does just that. We indicated in Chapter 1 that the goal of our search was a technical civilization, defined as one with the capability for interstellar radio communication (a more detailed discussion of which will appear in the next chapter). A civilization that sends radio messages into space could provide its own beacon, alerting other such civilizations to its existence and guiding their spaceships to its home planet. Perhaps the radio emissions from our terrestrial civilization have been picked up by aliens, who have followed the signals back here to Earth. This *could* be an explanation for the presence of aliens on Earth, but there are a few difficulties yet to be ironed out. Chief among these is the distance factor.

Radio waves travel at the speed of light. A signal emitted from the Earth will require 10 years to travel a distance of 10 light-years, 50 years to go 50 light-years, and so on. As we have been using radio waves for only about a century, our very earliest radio transmissions can be no more than about 100 light-years away from Earth by now. Alien civilizations living beyond this limit — and in a galaxy 80,000 light-years across, most of them *should* be farther than 100 light-years away — will have had no indications as yet that we exist as a technical civilization.

Furthermore, if we expect that aliens have already arrived here in response to our signals, they must be even closer, to allow for their own travel time. Assuming very superior civilizations that can travel at *nearly* the speed of light, we should expect that their travel time to Earth will be about the same as the time required for our radio signals to reach them. If they had their space ships all ready to go, just waiting for the first radio waves to arrive from Earth, then we must allocate half of our century of radio for each leg of the trip: 50 years for the radio waves to get there and another 50 years for the aliens to come here. If the aliens learned of our presence by way of our radio broadcasts, they must have come from no more than about 50 light-years away — a *tiny* portion of the Galaxy.

In reality, the situation is probably even worse. The very early radio signals used on Earth were neither beamed out into space nor powerful enough to be detected from nearby stars. And a civilization that *did* manage to pick up our signals would probably have a difficult time pinpointing their source and thus knowing exactly where to come. Whether they would be inclined to embark on a journey to an undetermined target is somewhat unclear (although it is done all the time in the movies).

MISSION MOTIVATION

Given the apparent difficulty of interstellar space travel, we might also wonder what *motivation* aliens would have in coming here. Just because a civilization has the technical ability to make such journeys does not mean that they have the resources to explore every corner of the Galaxy. Presumably they would have to carefully select their mission targets from a number of possibilities. What would make them choose Earth?

The aliens' knowledge of Earth should be fairly limited depending on how far away they are and how much information they have gathered from our radio signals. They might correctly identify the Sun as our star, but observing the Earth itself will be considerably less feasible. They may detect radio signals coming from the general direction of the Sun and thus deduce the presence of a technical civilization, possibly on one of the Sun's planets, but making any sense of the jumble of radio noise produced by all of the transmitters on Earth will be a formidable task. They may know little more than the fact that we exist and have discovered radio waves – not much ammunition upon which to base a mission. If that is the case, then the aliens will have little more than curiosity to drive them to come here (assuming curiosity is indeed an alien characteristic).

But how curious are they apt to be of *us*? If these aliens are far superior to us – which will probably be necessary if they are to travel between the stars – then they will not be likely to learn many technical tricks from us or advance their understanding of science very much. If space-faring technical civilizations are common in the Galaxy, then we are little more than fledglings with relatively little to offer. Our biology may be of some interest to their scientists, or possibly not, if it is not really all that unique. If we are just another carbon-based biological system – one of hundreds or thousands scattered throughout the Galaxy – then we may not be worth the time and effort it would take to come study us up close.

One could argue that perhaps we *are* unique, that there are very few technical civilizations around, and just meeting us face-to-face (if aliens have faces) would be enough motivation for them to come here. This is of course possible, but highly unlikely. If technical civilizations are, in fact, rare in the Galaxy, then the probability that one of them will actually be close enough to locate and visit us must be extremely small.

If we want to argue *for* the presence of extraterrestrials here on Earth, we have two different scenarios to consider: (1) space-faring technical civilizations are very abundant in the Galaxy, and there are several fairly nearby who decide to come here to study us, despite our being just barely technical and otherwise nothing special; (2) space-faring technical civilizations are extremely rare in the Galaxy, but there is one very close to us that comes here to visit. Although either of these *could* be true, neither of them seems particularly probable (due to the requirement that the aliens must be relatively close to us in either case). And the goal of science is to determine the *most likely* hypotheses for our observations.

THE UFO CONNECTION

Either extraterrestrial aliens are here or they are not. If they are *not* here, then *all* claims of alien sightings and encounters are false, being based instead on misinterpretations (of terrestrial, celestial, and mental phenomena), outright hoaxes, or lies. If aliens *are* here, *some* of these claims may indeed be true, but there is still room for plenty of misinterpretations, hoaxes, and lies. In fact, if they really have the power and the will to remain undetected, the aliens could be here, *and* all of the claims of sightings and encounters could be false! If UFOs and the like actually have *nothing* at all to do with extraterrestrials, then they cannot be used as evidence on *either* side of the debate.

If aliens came to Earth, what sort of evidence might we expect to see? If they made *no* attempt to hide or disguise themselves, their arrival should be obvious. Scattered around the world are a few thousand professional astronomers and even more amateur astronomers who observe the sky on a regular basis. Knowing the sky as well as they do they should be quick to notice anything so out of the ordinary as a starship approaching the Earth. Upon entering our atmosphere, such a vehicle should show up on the radars of the military installations in various nations, equipment designed specifically to detect intruders in the sky. Civilian airports, weather stations, and some of our orbiting satellites could also be expected to detect spacecraft in our atmosphere if indeed they are here. But while there have been occasional strange radar blips to deal with, there has been no confirmed report by these agencies of the arrival of extraterrestrial spaceships. Either the aliens are not here, or they are very good at avoiding detection.

In order to arrive here completely unseen, the aliens must have planned ahead, developing stealth spacecraft that are virtually invisible to all of our detectors. Of course, before their arrival, they would be unaware of what types of detectors we have, which might make their task trickier. It should be fairly obvious, however, that a craft that displays artificial lights while flying around in our atmosphere is going to attract some attention. Terrestrial aircraft show lights on the wings, tail, etc. to help them avoid collisions at night, but it would make little sense for secretive aliens to take the same precautions. Thus it seems highly unlikely that strange lights seen in our night sky are the artificial lights of alien craft.

There are a host of other explanations for lights in the sky. Civilian aircraft, military aircraft (some of which may be top secret and totally unknown to the general public), balloons, satellites, meteors, and planets are the most common objects involved. Some of these display their own running lights and/or landing lights when near the ground. Others, by virtue of their high altitudes, may simply reflect sunlight to the observer on the ground. Meteors are special because their light is generated through friction between meteoroids and the atmosphere as these tiny bits of rock make their attempts to land on Earth.

Lights in the sky may be perceived to be moving or still, but motion can be deceptively difficult to judge. An object moving directly toward the observer may appear to be stationary even though it is actually moving very rapidly. A change in course may cause the object to *seem* to suddenly acquire a very high speed and appear most remarkable to the casual observer. Strangely, a stationary object may sometimes be perceived as moving if that makes more sense to the observer's brain. There are tales of World War II pilots being chased by a persistent bright light that countered every evasive move they made and could not be eluded. Luckily, it did them no harm either: the planet Venus carries no weapons. Similarly, the twinkling and flashing colors of bright stars and planets seen close to the horizon can give the illusion of rotation to the casual observer.

Velocities are very difficult to judge without knowledge of the object's distance. Our normal cues for distance usually involve an object's apparent size, but at night, with only a light to go on, size is of little use to the brain. An object that appears to move at a tremendous speed may actually be much closer than the observer believes. Even in the daylight when the apparent size of an unidentified object can be perceived, without knowledge of its *actual* size, its distance – and its actual velocity – will be undetermined.

There are many ways to explain observations of UFOs without resorting to extraterrestrial spaceships. Of course, with all the different sightings, there is no *single* explanation that covers every case. And in many cases, there is insufficient observational evidence to identify which of several alternative hypotheses is best. Space travel appears so difficult, so perilous, and so time-consuming that it should not be invoked as a solution to UFO sightings without careful consideration of all its implications and review of the more mundane terrestrial explanations.

RELATED TALES

If UFOs can be explained without resorting to alien spaceships, then it is most likely that extra-terrestrials are not here after all. Of course, there are other observations that have been interpreted as evidence of an alien presence; how can alien abduction stories, the Roswell incident, and the like be explained without invoking extraterrestrials?

Alien Abductions

A number of people claim to have had contact with aliens, with some reportedly having been taken aboard spaceships and subjected to physical examinations. The descriptions of aliens offered by different abductees are often very similar to each other, which would seem to lend credence to their stories. After all, how could so many individuals have the same image of an extraterrestrial in each of their brains unless they had all encountered the same alien race?

The simplest response is that with plenty of alien images in magazines, movies, television, and the Internet, the general public has been thoroughly exposed to various perceptions of an alien's appearance. Practically anyone – even those making no claim of alien encounters – can describe an intelligent alien being. Such descriptions frequently involve a fairly humanoid body with two arms, two legs, and a head; the face normally has two eyes and usually a mouth and/or nose – all arranged in the same approximate positions as on a human head.

As evidence, consider an experiment involving students in the author's *Life in the Universe* course who were asked (fairly early in the course) to sketch "an intelligent alien". Of the several hundred such sketches made over a number of years, 71% of the alien *bodies* drawn showed a humanoid pattern, while 76% of the alien *faces* sketched were ranked as humanoid (as described in the previous paragraph). None of these students claimed any personal contact with aliens and yet their alien descriptions exhibited a general humanoid theme. Apparently we humans have a strong tendency to visualize intelligent aliens as being similar to ourselves; or perhaps we just lack sufficient imagination to envision realistic beings that might have a significantly different body plan.

Should intelligent aliens have humanoid bodies? Is there something special about the humanoid form that makes it the best choice for intelligent species in technical civilizations that travel to the stars? One can make arguments about the design of our bodies – the advantages of walking on two appendages rather than four, freeing the hands to do other interesting things; the position of the eyes at the top of the erect body where they enjoy a maximum range of vision; the location of the brain inside a protective skull and close to the major sensory organs; etc. – as being superior for an intelligent species, the pinnacle of evolution. That may be true on the Earth, but extraterrestrial evolution will not necessarily stumble onto this same body plan as it develops an intelligent alien species. Most of the creatures on Earth look nothing at all like humans, despite having the same biochemical basis and a common origin in the distant past. We should not expect aliens to resemble us, at least not in very many cases. That humanoid aliens not only exist, but also live close enough to Earth to come here would be a most remarkable coincidence.

But perhaps this low-probability event has in fact occurred and humanoid aliens are here. Or perhaps there is indeed something about evolution that drives the production of intelligent humanoids and essentially guarantees that visiting aliens will resemble us (thus increasing the demand for human actors in science fiction movies). If either of these is the case, then why would these aliens be abducting humans and then *returning* them? If they were really interested in understanding our biology, it would be much simpler to abduct several of them, dissecting some and keeping the others for future study. If they were really concerned about the lives of the individual abductees, they could tap into the Internet or steal some biology books from the library instead. If they were really trying to keep their presence here a secret, they would certainly not want to return the abductees to their homes to let some hypnotist pry the secret from their subconscious memories. These are supposed to be *intelligent* aliens.

We are all capable of dreaming – letting the brain spin stories for us while the conscious mind is idling or at rest. Sometimes these dreams have strong connections to reality but many times they do not. Sometimes we can almost believe what the brain is telling us in these dreams, even after waking up. One explanation for tales of alien abductions is that they are concocted by an individual's brain – possibly with the assistance of a hypnotist – with enough familiar images to make them appear real. An individual who is predisposed to believe in aliens may well have a difficult time denying such a story and may ultimately accept it as the truth. In presenting their personal tales of alien abduction, these individuals are not lying or just spinning yarns but rather, recounting stories that they believe to be true. And this sincerity makes their tales appear more credible. Which is easier to believe – undetectable aliens traveling light-years through space to study humans in person, or figments of the imaginations of some individuals' brains? The choice of these – or other plausible hypotheses – is left to the reader.

Roswell

Another interesting saga is the now-famous Roswell incident, which began back in July 1947 near Roswell, New Mexico, when something apparently crashed in the desert. A rancher found the wreckage and subsequently notified the local Air Force base. The military dispatched crews who gathered up the wreckage, but not before a number of citizens had caught a glimpse of some of the bits and pieces, which appeared to be composed of unfamiliar materials.

One of the initial explanations given by the military was that a weather balloon had crashed, which would not have been an unusual event. However, the swiftness and secrecy with which the military moved to collect all evidence of the crash caused some people to doubt the story. The strange materials seen by some of the onlookers led to the fantastic tale of an extraterrestrial spacecraft having crashed in the New Mexico desert, an explanation apparently more in keeping with the frantic response of the military. (Why get so excited over a weather balloon?) Thus, a legend was born.

Today many people are convinced that this explanation is correct – that an extraterrestrial spacecraft crashed, the military collected and hid the evidence, and the government has been keeping the whole affair secret ever since. This viewpoint has persisted because the military and the government *do* have secrets that are not made public, thus making that part of the story very plausible. The government does not regularly divulge every scrap of information it has because it does not want potential enemies to gain an advantage over us. The military does not advertise the characteristics of every new weapon it adopts because it does not want potential enemies to have time to develop countermeasures.

It would be quite fitting that these two entities would want to keep something as spectacular as a crashed spacecraft a secret. The military might be able to learn about some highly advanced propulsion system or superior construction materials, possibly copying and employing them in future aircraft. The government might gain some leverage to use against its opponents in the cold war. But the fact that neither of these things appears to have occurred could lead one to doubt the alien spacecraft theory.

In the 1990s the military produced another explanation that is consistent with the observations. (It was, of course, rejected by those who believe in the alien spacecraft theory, who feel that *anything* the government or military says about Roswell is just part of the continuing cover-up.) In this version of the story, the crashed object was a military balloon, being test flown as part of a secret project designed to monitor certain activities of our enemies. Understandably, the military did not want to report the *truth* about the crash to the local news media, nor could they seem to agree on a cover story – which explains both their secrecy and their initial confusion. The linking of the crash to an alien spacecraft actually worked out fairly well for them as they could easily – and truthfully – deny any knowledge of the aliens' origins or intentions.

With most of the original participants in the Roswell incident no longer living, we are not apt to obtain any definitive testimony that might clarify what actually happened. Once again, the reader is left to choose between two main hypotheses: either the aliens managed to travel across the vastness of interstellar space only to splatter their spaceship into the sand near a New Mexico air base, *or* the military suffered a minor setback in one of their secret projects and did not wish the public to find out any details about it.

Crop Circles

Beginning some time in the 1980s, strange designs began to appear overnight in the wheat fields of southern England. Being basically circular in shape, they were termed **crop circles**. Since then, the crop circles have increased in number and complexity and spread throughout the grain-producing areas of the world, although England continues to be the principal focus of these activities.

The origin and purpose of the crop circles were not immediately obvious. Some who analyzed the patterns concluded that they could not possibly have been made by humans, especially in the small amount of darkness available on summer nights in England. The circular nature of these flattened crops led some to speculate that they had been made by spacecraft landing in the fields at night, while others attributed them to some mysterious, unknown forces of nature.

In the 1990s, two Englishmen came forth to explain that *they* had begun the crop circle craze, using little more than a rope and a board to smash down the stalks of grain. Although they willingly demonstrated their technique, not everyone was ready to believe that humans could have performed such a feat. But hundreds of crop circles have continued to appear, as more and more crop circle makers take up this challenging hobby. Head-to-head contests have even been held to determine which team can make the *best* crop circles (whatever that means). In this case it would appear that there is little need for extraterrestrials to come here to make crop circles as the locals seem to have the project well in hand.

Ancient Astronauts

In a similar vein, there are many *ancient* structures that appear so massive and/or complicated that some people cannot conceive of their construction by the human civilizations that would have been in existence at the time – at least, not without outside help. In the 'ancient astronauts' theory, extraterrestrials visited Earth in the not-too-distant past and assisted, encouraged, and/or inspired the building of the pyramids of Egypt, the carving of the stone heads on Easter Island, the drawing of the huge animals in the Nasca desert in Peru, etc. Are these wonders evidence of a previous extraterrestrial presence on Earth?

Two points come to mind. First, construction of pyramids and the like by early human civilizations no doubt required huge investments in time and resources, along with a fair degree of technical skill. But without television or the Internet, without warlike neighbors to defend against, a peace-loving society may have welcomed the mental and physical challenges of such creative efforts, and wise rulers would probably have been very glad to have their people engaged in such long-term activities. In short, these may have been very appropriate projects for the civilizations of the time to tackle, especially if they served some political, religious, or artistic purposes as well.

Second, it seems very odd that the aliens would have had the foresight to bring along some-one sufficiently skilled in working with *stone* to be able to give advice to people who used it constantly in everyday life. Although we cannot say for sure, it certainly seems reasonable to expect that extraterrestrial spacecraft will *not* require the services of an in-flight stonemason in order to keep them in working order. A monument that required advanced knowledge and skills in electronics, computers, nuclear physics, or metallurgy (or some other area so exotic we have not yet discovered it) would be much more convincing to us and even *more* impressive to the ancient civilization. If the aliens meant to provide human civilizations with a technology upgrade, they could have done a far better job; if they merely intended to leave a calling card, they could have made it much more obvious.

FIT FOR TRAVEL

Humans evolved on Earth over the last few hundred thousand years or so, with bodies specifically adapted to this terrestrial environment. But while humans have developed and thrived in a number of different climates with considerable variation in temperature, humidity, and air pressure, they have always had a constant gravitational field. Only recently, as astronauts began to venture into space, has anyone managed to temporarily escape the clutches of the Earth's gravity.

Space travelers on interstellar missions will experience no sensation of weight, unless some artificial gravity is supplied by constant acceleration of the spacecraft. In this weightless condition, the astronauts' bodies will change. Bones will gradually lose mass without the constant stress applied to them to hold up the body; leg muscles will weaken through lack of use; and the heart, which will have a considerably easier task pumping blood around through a 'weightless' body will also adapt to a less strenuous routine. The longer the journey, the greater the effects will be; but the real problem will occur at the end of the trip, when the astronauts again find themselves in the substantial gravitational field of the planet on which they land.

Readjustment to a once-familiar gravity or adaptation to a completely new field will not occur immediately. We should not expect our returning astronauts to come bounding down out of their space vehicle after their trip to the stars, ready to march in a celebratory parade. Although we have some data on the effect of several months of weightlessness on human bodies, we really have no experience with trips lasting several years. These unknown long-term effects on our bodies must be confronted before we attempt to travel as far away as Mars. It would make little sense – and be extremely embarrassing – to send astronauts on a distant voyage only to have them unable to walk when they arrive at their destination.

Just as we must take precautions to insure that the bodies of our astronauts do not deteriorate during long space voyages, any alien visitors to Earth will have had to do the same in order to remain in good physical condition during their long trip here. We do not know how easily alien bodies will adapt, either to weightlessness or to Earth's gravity and atmosphere, but we should probably not expect aliens to be particularly spry as they land on the Earth's surface for the first time. Perhaps those abductees who report seeing a fitness center on board the aliens' spacecraft should be taken more seriously.

All things considered, the case for extraterrestrials visiting the Earth appears weak at best. But that does not mean that aliens do not exist; they might be staying at home (as we are), just waiting for the phone to ring. Better still, they might be trying to get in touch with us right now! But how would they do that? How can we possibly make contact with aliens we have never met, given that tens, hundreds, or thousands of light-years of space separate us? In Chapter 9 we will explore some possibilities.

MAIN IDEAS

- The distances between stars are so great as to make interstellar space travel extremely difficult, if not impossible for beings with lifetimes similar to ours.

- Special relativity offers some possible solutions (time dilation and length contraction) that might allow interstellar journeys to be made within a human lifetime; even with special relativity, there are a number of problems that must be overcome before interstellar travel can become practical.

- Only within the last century has our technical civilization had the means to alert nearby aliens to our presence; however, only aliens living within a few dozen light-years of us could have had time to visit Earth following receipt of such information.

- Despite a host of reports of unidentified flying objects and numerous claims of abductions of humans by aliens, the scientific community has found little evidence to support the hypothesis that intelligent beings from another world have arrived at our planet.

- Given the difficulties of space travel, the vastness of the Galaxy, and the inconspicuous nature of the Sun and Earth, it would seem highly improbable that aliens are visiting our planet with sufficient frequency to account for even a tiny fraction of the UFO sightings.

KEYWORDS

alien abductions	ancient astronauts	crop circles
Doppler shift	length contraction	proper length
proper time	reference frame	rest mass
Roswell	special relativity	time dilation
unidentified flying object (UFO)		

LAUNCHPADS

1. Suppose that human lifetimes could be increased to 10,000 years. What effect would this change have on our plans for interstellar space travel?

2. If space travelers were to leave Earth now on a relativistic journey to the stars, they might come back here many decades or centuries in our future. What will be the public reaction to their return?

3. Could *all* UFO sightings actually be the result of alien visits to Earth?

Chapter 9
MESSAGES IN SPACE

*I*n *which properties of electromagnetic waves are described, the electromagnetic spectrum is presented, optimal frequencies for sending and receiving interstellar messages are discussed, other factors affecting SETI searches are examined, binary code is explained, and procedures for sending and decoding pictogram messages are illustrated.*

It would appear that space travel is very difficult – perhaps so difficult that *no* civilization will be able to master it. If that is the case, we will never be able to meet any aliens 'in person', which may be a good thing, depending on their temperament. But we might be able to meet their *minds* and converse with them, exchanging ideas, customs, and scientific knowledge about our respective corners of the Galaxy. All we need is a means of communicating across interstellar distances.

We could send automated space probes. With no life-support systems, the payload would be much smaller and simpler, and the whole spacecraft would be less expensive to construct and launch. We could send out numerous probes for the cost of a single manned mission and not have to worry nearly so much about the perils of space. This tactic has already been successfully utilized to explore much of our solar system, with probes in the Mariner, Pioneer, Viking, and Voyager series providing coverage of most of the planets.

Upon completion of their missions, these probes either crashed/landed on a planetary surface, went into orbit about a planet or the Sun, or left the solar system to sail into deep space. This last fate has been shared by most of the probes to the outer planets, which were propelled out away from the Sun with sufficient speed to escape its gravitational pull. Launched in the 1970s, spacecraft such as Pioneer 10 and Voyager 2 are not headed toward any particular star, nor are they traveling at relativistic speeds. They are destined to drift slowly among the stars, unlikely to be found by anyone – each the interstellar equivalent of a message in a bottle.

Just in case, the Pioneer 10 probe carries a plaque with sketches of itself and two humans, a diagram of the solar system, and a plot intended to depict our position in the Galaxy relative to several pulsars (rotating neutron stars that emit pulsed radiation). Similarly, the Voyager spacecraft each carry a long-playing phonograph record (pre-compact-disk technology) containing sounds of the Earth – human speech and music, sounds of nature, etc. – and electronically encoded images of various terrestrial scenes. Although a record *player* was *not* included in the package, instructions for building one were; but whether the lucky aliens who find a Voyager will ever actually view and hear the message is debatable.

Obviously, there is a need for a more efficient means of communicating between stars. As with attempts at space travel, the root of the problem is still the vast interstellar distances, and once again, the solution is to send something at the highest possible speed. This time however, we need not send material objects; messages can be transmitted quite well in the form of radiation, such as light. Because all radiation of this type travels at the speed of light, this is also the most rapid means of sending a message.

ELECTROMAGNETIC WAVES

Light is an interesting phenomenon, but it is somewhat elusive to define. In some experiments light behaves as if it is made up of tiny *particles* – little bundles of energy called **photons** that travel from one point to another. In other experiments, light exhibits the properties of a *wave*, traveling through space from a light source to the observer. It is the wave nature of light that will be described here.

Light is one form of **electromagnetic radiation**. Matter and space contain electric and magnetic fields; the periodic *fluctuations* in these fields that travel through matter and space make up electromagnetic radiation.

When describing this radiation as a wave, we make use of the general properties of waves. The simplest to understand is the **wavelength** (λ = *lambda*): the distance between crests of a wave. A second property is the **speed** at which the wave travels. In a vacuum, this speed is a constant (c) for *all* electromagnetic waves: c = the speed of light – about 300,000 kilometers per second. Wavelength and speed are illustrated in Figure 9.1.

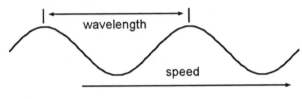

Figure 9.1: Wave properties.

The next wave property is not so easily illustrated. If we stand at one location in the path of the wave and count the number of wavelengths that pass by every second, we are measuring the **frequency** (ν = *nu*) of the wave, measured in cycles per second (or hertz). Wavelength, frequency and speed are related by a simple equation: $\lambda\nu$ = c.

Because c is a constant, the wavelength and frequency are not independent of each other: if the wavelength is long, the frequency will be low, and if the wavelength is short, the frequency will be high. If one is known, the other can be calculated. Therefore, we can identify different electromagnetic waves using *either* the frequency or the wavelength.

Electromagnetic waves have an additional property of interest: **energy** is carried by the wave from the source to the observer. The energy of a wave is proportional to its frequency (E = hν, where h is a constant). Thus a high frequency wave has high energy and a short wavelength, while a low frequency wave has a low energy and a long wavelength. Figure 9.2 illustrates this result.

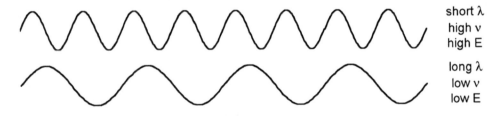

short λ
high ν
high E

long λ
low ν
low E

Figure 9.2: Wavelength, frequency, and energy relations.

The range of wavelengths available for electromagnetic radiation is essentially unlimited. Various natural and artificial processes produce radiation at different wavelengths, and suitable detectors must be employed that will respond to the radiation of interest. The entire range of wavelengths (or frequencies or energies) is called the **electromagnetic spectrum**; we divide the spectrum up into different wavelength regions as shown in Figure 9.3.

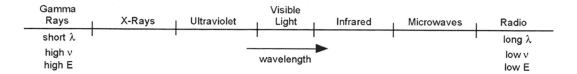

Figure 9.3: The electromagnetic spectrum.

Visible light comprises a relatively narrow band of wavelengths that can be detected by the human eye. At shorter wavelengths is ultraviolet radiation; the higher energies associated with these rays can cause sunburn when absorbed by the skin. The even higher energies of X-rays allow them to penetrate the skin, permitting imaging of a person's skeleton. Gamma rays have the greatest penetrating power, which makes them very useful in the treatment of tumors.

At wavelengths somewhat longer than those of visible light are infrared rays; these are invisible to the eye, but they can be absorbed by the skin as heat rays. Microwaves have frequencies that match the vibrational frequencies of many molecules, including water; irradiating substances containing water molecules with microwaves of the right wavelength causes the water molecules to absorb energy, the principle behind microwave ovens. Radio waves have the longest wavelengths and lowest energies of all; they are not generally detectable by human bodies but are the basis of a variety of electronic gear, including television, AM radio, FM radio, cell phones, short-wave radio, etc. (Note: In some cases, microwaves are included in the general category of radio waves.)

Another common type of wave does not appear in this list. Sound waves are *not* a form of electromagnetic radiation; instead, they are fluctuations in the pressure or density of the matter through which they pass. When you listen to a radio, your ear is detecting sound waves created by the speaker in response to electrical signals from the amplifier; these signals came to the amplifier from the antenna, which created them in harmony with the (electromagnetic) radio waves it absorbed from the transmitting tower. Because sound waves require the presence of matter for their transmission and because the speed of sound is considerably less than the speed of light, sound waves will not prove very useful for transmitting messages across the vacuum of space.

SELECTING THE MESSENGER

Electromagnetic radiation is a fine choice for sending interstellar messages because it can travel across a vacuum and moves at the ultimate speed – the speed of light. *All* types of electromagnetic radiation share these two properties, but they vary in the ways they interact with matter. This variation can help to determine which type of electromagnetic radiation should be the most logical choice for transmitting and receiving messages.

Suppose we wish to audition different types of electromagnetic radiation for the role of interstellar messenger; what requirements will the winning candidate have to meet?

We may envision beaming electromagnetic radiation toward a particular star system, to an alien civilization that may reside there. The beam, generated by suitable equipment on Earth, would first have to pass through Earth's atmosphere before reaching the vacuum of space. Then it would travel through space to the aliens' planet, where it would again have to pass through an atmosphere and be received by appropriate detectors on the planet's surface. Clearly, radiation that is severely attenuated by passage through our atmosphere will not be especially useful in this endeavor; this criterion alone would restrict our choices to two spectral regions: visible light and some of the radio/microwave wavelengths. (While the atmosphere is somewhat transparent to infrared and ultraviolet wavelengths, we would be much better off using the more transparent visible and radio/microwave windows.) Although the aliens' atmosphere may be completely unknown to us, we can speculate that it will be somewhat similar to ours in its transmission properties, giving the same two windows in the spectrum. (This should be a fairly safe bet, given the biochemical similarities we expect in our life forms and theirs.)

It could be argued that an intelligent civilization will have the capability and the desire to position its receivers and transmitters high above the atmosphere aboard orbiting satellites or on the surface of some barren world such as the Moon or an asteroid. In this way, the atmospheric requirement can be avoided, and *any* wavelength can be used. Although *we* are not yet at this stage of technical development, this point should be kept in mind for the future.

Another requirement can be placed on the radiation by its passage through space. We have already noted that space contains matter in the form of interstellar clouds of gas and dust. These clouds can inhibit incident radiation by absorbing and/or scattering it, which could be fairly disruptive to an interstellar signal. The ideal choice for a message would be radiation that is not affected by the interstellar medium in any way. Visible light is not a particularly good candidate as it is easily absorbed by dense interstellar clouds, preventing us from seeing very distant stars in the plane of the Milky Way. The best selections on this basis would probably be the two extremes of the spectrum – gamma rays and radio waves – as their energies are least likely to match those required for interactions with normal matter.

Getting there is only part of the difficulty facing the interstellar messenger wave; in order for a civilization to be able to utilize it, the radiation must clear a few more hurdles. First, of course, the civilization must have *discovered* the chosen type of radiation, along with some means of producing and detecting it. Ideally, this discovery should not take too long or the civilization's 'technical' rating will be delayed. A form of radiation that requires minimal – or at least, easy and obvious – technology to utilize will be the most widespread among the technical civilizations of the Galaxy. More advanced civilizations may choose to adopt trickier technologies for their communications, but anyone attempting to contact entry-level civilizations will want to keep things fairly simple.

Of the different forms of radiation, the most obvious to us must be visible light, which humans have been using as long as they have had eyes. Astronomy was originally based on observations in the visible range because no special technology was required beyond the eyes of the observer. Development of telescopes enhanced our view of the lights in the night sky, but most of the spectrum remained unavailable for astronomical use, due to the effects of our fine atmospheric filter. Astronomical observations are now performed in all of the major regions of the spectrum, but most of these wavelengths had to await the development of rockets and satellites before becoming practical observational tools. However, radio astronomy – which can be done from the surface of the Earth – preceded satellite technology by several decades and thus could be considered second to visible astronomy in this category. If alien civilizations exist in atmospheres similar to ours, and their technology proceeds at a rate similar to ours, the radio/microwave window should be one of the first new regions of the spectrum that they will explore.

A civilization that intends to send lots of messages to the stars must be prepared to pay for them. Apart from the cost of technology and any administrative fees, the major expense for these messages may be the energy that is beamed out as electromagnetic waves. Radiation that requires little energy to produce will provide a more efficient use of resources and permit a greater amount of communication for the alien civilization on a tight budget. The form of electromagnetic radiation with the lowest energy requirements is, of course, radio waves.

At this point, a brief inspection should indicate that the best candidate for the job might well be radio waves. These pass in and out of our atmosphere (and presumably, the aliens') with relative ease, are not seriously affected by the interstellar medium, should be easily discovered and used for astronomical observations by developing civilizations, and have minimal energy requirements to boot. What could be better?

There is an additional problem to consider. As already noted, there are many astrophysical sources of radiation, covering a large portion of the spectrum. Stars produce ultraviolet, infrared, and visible light; planets are infrared sources; nebulae, pulsars, and galaxies generate radio waves; very hot gases swirling around black holes can radiate X-rays; colliding neutron stars spray out bursts of gamma rays, etc. Not all of this radiation is generated in equal amounts, with some parts of the electromagnetic spectrum being noisier than others.

Artificial noise must also be considered, such as that produced by the equipment used to detect radio signals. An amplifier hum may be heard when a stereo volume control is turned up too high with no music playing; similarly, the electronic components used to amplify the faint radio signals from space will introduce their own noise into the result, making the signal from the source more difficult to detect. The power level of noise across the spectrum is commonly specified by the **noise temperature**. (This term is used because the power level in the radio region of the spectrum is proportional to the temperature of the radiating body.) Equipment with a higher noise temperature produces more noise in the signal, which would appear as a louder hum or hiss in the radio speaker if we were listening to a cosmic radio program.

Noise is not distributed uniformly across the spectrum; some kinds of noise are more prevalent at high frequencies while others are more apparent at low frequencies. Figure 9.4 illustrates the spectral distribution of the primary sources of noise: amplifiers, the Big Bang, and the Milky Way Galaxy itself.

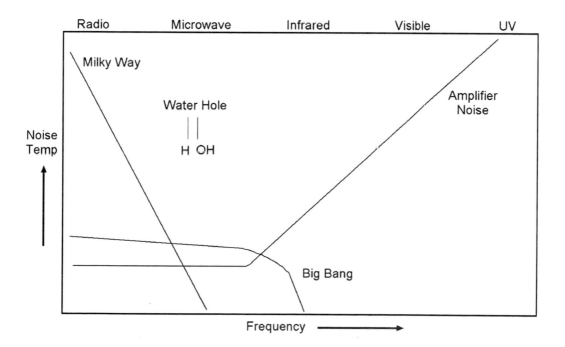

Figure 9.4: Noise levels across the spectrum.

A civilization attempting to make itself heard will want to utilize radiation that stands out from the natural noise of the cosmos, rather than being lost in the din of electromagnetic waves bombarding their planet. At the same time, it is undesirable for a low power signal to be overwhelmed by the noise level of the amplifier being used. These two constraints tend to eliminate the extremes of the spectrum and direct the choice toward the radio/microwave frequencies. Happily, this selection is in agreement with the result of our first set of criteria.

THE WATER HOLE

Limiting messages to the radio/microwave region of the spectrum does not exactly solve the problem, just as the knowledge that your cousin lives somewhere in the state of California does not exactly tell you where to find her. A radio telescope (or a microwave antenna) operates at a particular **frequency**, receiving only those signals with frequencies at or very close to this operating frequency. In order for an extraterrestrial message to be picked up, the receiver must be tuned to the same frequency as the transmitting radio telescope, just as your car radio must be tuned to the correct frequency in order to receive your favorite station. Each region of the spectrum includes a vast range of frequencies to search for possible messages, making this a nontrivial exercise.

Ideally there would be a preset frequency to use – one upon which all civilizations could agree. Of course, the logistics of making such an agreement without first establishing lines of communication would be a bit tricky; and if communications *could* be established without knowing this 'magic' frequency, there would be much less need for trying to agree on one. As it turns out, there just may be a magic frequency that becomes obvious to all civilizations that decide to send or receive interstellar messages.

That the most abundant atom in the universe is hydrogen will be known to all such civilizations. The neutral hydrogen atom radiates at a characteristic frequency of 1420 MHz (1 MHz = one megahertz = 1 million cycles per second), producing a wavelength of 21 centimeters in the vacuum of space. This radiation has been used by radio astronomers to map the positions of interstellar hydrogen clouds in the Galaxy. Alien civilizations may well want to use a frequency close to 1420 MHz because they will know that other civilizations are probably doing radio astronomy in this same spectral region and will be more apt to find a signal there.

We can take this one step further by investigating other atomic and molecular frequencies in the microwave band. Around 1662 MHz – close to the atomic hydrogen emission at 1420 MHz – is a group of frequencies for the hydroxyl radical (OH). Because H and OH combine to make H_2O, and because water is an essential substance for terrestrial life, we earthlings find this range of frequencies between 1420 and 1662 MHz – dubbed the **water hole** – to be particularly intriguing. If the extraterrestrials are also made of water, perhaps they too will be attracted to this spectral region. And if they have deserts on their planets with real water holes where different life forms gather to communicate with each other, perhaps they will find these frequencies especially significant. The location of the water hole in the valley of relatively low noise in Figure 9.4 makes these frequencies very special indeed.

A MULTIDIMENSIONAL PUZZLE

Now it might seem that knowing which frequencies are most likely to be used for interstellar messages would provide us with sufficient information to insure success, but this is not necessarily so. There are several other parameters that must also be matched before contact can be made.

For starters, there is the **Doppler shift**, mentioned in Chapter 8. This is the apparent shift in the frequency (or wavelength) of a signal due to relative motion between the source and the observer. When we observe any star from the Earth, the star's light will be shifted by the rotation of the Earth on its axis, the revolution of the Earth about the Sun, and the motion of the star with respect to the Sun. For a signal originating on a planet about the star, the rotation and revolution of the planet must also be included. Thus, if the aliens send a signal at 1420 MHz, we will have to tune our equipment slightly above or below this value to receive the message at the Doppler-shifted frequency. Additional messages sent out from the same planet at a given frequency will not necessarily all arrive here at the same frequency, due to changes in the relative motion over time. Although the overall effect of the Doppler shift should be fairly small for stars within our Galaxy, it must be addressed if very precise frequencies are desired.

Most detectors respond to a *range* of frequencies, rather than just one. The human eye is sensitive to the many different wavelengths of visible light, which the brain interprets as different colors. Astronomers often use filters on their telescopes to restrict the light received to a narrower range of wavelengths. A radio telescope tuned to a particular frequency is still sensitive to slightly higher or lower frequencies but essentially blind to frequencies much beyond these, just as an AM or FM radio ignores signals from stations other than the one to which it is tuned. The range of frequencies to which a detector responds is called the **bandwidth**. Similarly, a message sent by a transmitter tuned to a particular frequency

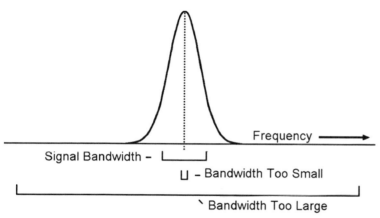

Figure 9.5. Bandwidth matching.

will actually be spread over a range of frequencies determined by the transmitter's bandwidth.

If we match our detector's frequency to the frequency of an incoming message, our reception of the message will still depend on the relation of our detector's bandwidth to that of the transmitter. Figure 9.5 shows a signal centered on a particular frequency (vertical dotted line), with its bandwidth marked by the bracket just below. A detector tuned to the same frequency and set to approximately the same bandwidth should be able to receive the signal.

However, if the detector's bandwidth is much *smaller* than that of the signal, only a small portion of the signal will be received, with most of it going undetected. On the other hand, if the detector's bandwidth is much *larger* than that of the signal, too much noise from frequencies outside the signal's bandwidth will be picked up, and this noise will make the signal more difficult to identify. For proper reception, the bandwidths of the signal and the receiver should be comparable.

Another criterion to be met involves the choice of an antenna. Most commercial radio stations use an antenna that broadcasts a signal in *all* directions along the Earth's surface, in an effort to reach as many listeners as possible. Most automobile radios employ a similar antenna that receives radio waves from *any* direction around the compass, permitting the listener to choose from a variety of stations. Such an omni-directional antenna *could* be used to search for signals from aliens, but this plan has a major weakness. Although an extraterrestrial signal could be received without any special antenna alignment needed, the source of the signal would not be known; we would know that aliens exist but would have no idea where they might be. Additionally, the relatively weak signal would be mixed with noise received from all directions, making it far less likely to be detected.

A better plan is to use a radio telescope to point the antenna in a particular direction. The antenna will then pick up only the signals and noise coming from the small area covered by the beam of the radio telescope. Larger radio telescopes will be more sensitive to faint signals, just as larger optical telescopes make stars appear brighter. Use of the radio telescope allows us to determine the location of the source of the alien signal and increases our chance of detecting faint signals.

Of course there is a catch. Without knowing just *where* the aliens live, we will not know where to point our radio telescopes. Not only must we set our radio telescopes to the right frequency and bandwidth, but we must also point them in the right *direction* in order to receive a message.

A related problem involves *time*; we must be listening for the message when it is actually arriving here, and we must listen long enough to obtain the entire message. We could have our radio telescope set to the proper frequency and bandwidth and pointed in the right direction, but if the alien message is not arriving just then, we will not receive it.

There are several reasons why this could happen. First, the aliens are unlikely to mount an effort to *continuously* transmit radio messages toward our star, even if they suspect that we might be here. If they are trying to make their presence known, they will probably be spreading their efforts among several likely targets. Second, if they live on a rotating planet such as ours, its spinning motion may periodically carry our Sun below their horizon and out of radio contact for a while. Similarly, the Earth's rotation makes most of the stars we see appear to rise and set, with some of them always remaining below the horizon for observers at certain locations. Third, as radio telescopes can be used for *either* transmitting or receiving, the aliens may wish to occasionally stop sending messages and listen for a response.

Now if all of these conditions are met – the right frequency, bandwidth, direction, and time – we should be able to receive interstellar messages, unless the aliens are so far away that the strength of their signal is too low to rise above the noise level. The loudness of a sound, the brightness of a star, and the strength of a radio beam all diminish with increasing *distance* from the source. The effect is not insignificant; increasing the distance by a factor of 10 will decrease the signal strength by a factor of 100. If an alien civilization is too far away, its signals will be too weak to stand out from the background noise and will thus go undetected. Increased time spent observing a particular target can provide a higher signal-to-noise ratio, but less time will then be available for monitoring other potential sources.

Many of these difficulties can be addressed by a civilization in search of extraterrestrial communications. Multi-channel receivers can be built that will be capable of scanning many different frequencies at the same time; larger radio telescopes can be built to collect more signal from weaker and/or more distant sources; and more radio telescopes can be diverted from their normal research activities to listen around the clock for alien messages from neighboring stars. All that is required is the willingness to commit resources to the project.

If resources are not a problem, if a civilization has been able to adequately feed, house, clothe, and protect its citizens, and if it is willing to support its scientists in searching for messages from space, then such searches may certainly occur. However, if a planetary society is split by numerous political factions, if a significant portion of the population does not enjoy a reasonable standard of living, if the scientific community is compelled to research only those ideas that are deemed likely to yield positive results, then searching for potentially nonexistent extraterrestrial signals may be an unaffordable luxury.

SEARCHING FOR EXTRATERRESTRIALS

With the development of radio astronomy in the middle of the 20th century came the feasibility of actually searching for signals from extraterrestrials. But such attempts were not high on the initial list of research to be done; there were plenty of natural radio sources to be discovered, cataloged, studied, and explained by radio astronomers, few of whom were thinking about aliens.

The first search for radio signals from an extraterrestrial civilization was made in 1960 by Frank Drake. Dubbed Project Ozma, the search involved pointing the 85-foot-diameter radio telescope at Green Bank, West Virginia toward two nearby stars and listening for signals at frequencies around the neutral hydrogen line at 1420 MHz. Neither Epsilon Eridani (type K2, 10.7 light-years away) nor Tau Ceti (type G8, 11.8 light-years away) yielded any positive results in the approximately 150 hours of observing time devoted to the project (Breuer 1982).

Dozens of subsequent searches by other research teams, using various frequencies, additional stars, and different observing strategies have met with similar results. Whether targeting nearby stars, surveying stars in a particular part of the Galaxy, or simply scanning the sky for signals, researchers have still found no definitive evidence of extraterrestrial civilizations. There have occasionally been strange signals received that *could* have been produced by intelligent beings, but none has been confirmed as an extraterrestrial communication. Strange signals are quite understandable: our current society makes increasingly heavy use of electronic equipment, and it is difficult to find a region of the radio/microwave spectrum where contamination by terrestrial sources is not a problem.

As noted above, increased support for **SETI** (the **S**earch for **E**xtra**T**errestrial **I**ntelligence) radio search programs might eventually turn up some positive evidence – but only if alien civilizations are indeed transmitting signals in our direction. However, the lack of positive results has led many – including those in charge of distributing public funds for astronomical research – to believe that searching for aliens is a waste of time and resources and therefore should not be supported by taxpayers' money. This is not an unreasonable stance, for scientific research grants have generally been awarded to those projects that, in addition to pursuing questions of interest and importance, also show a high probability of success. Unfortunately for SETI proponents, the only way to demonstrate such a probability would be to locate an extraterrestrial civilization, and the best way to do that may well be through searches, which of course require funding: no funding – no searches; no searches – no positive results; no positive results – no funding.

At present, SETI observing programs are not particularly abundant. The principal current project, **SERENDIP** (**S**earch for **E**xtraterrestrial **R**adio **E**missions from **N**earby **D**eveloped **I**ntelligent **P**opulations), is a continuing effort run by the University of California at Berkeley. The SERENDIP equipment is mounted piggyback on the 1000-foot-diameter Arecibo radio telescope in Puerto Rico, the largest radio dish on the planet, where it scans whatever portion of the sky is being observed by other researchers. Due to the non-steerable nature of the Arecibo dish, this search is limited to about 10% of the sky.

The SETI@home program, begun in the late 1990s, has used millions of personal computers from around the world to analyze data collected by the SERENDIP equipment on the Arecibo radio telescope. This unique computing project provides individual computer users with free software that looks for particular patterns in incoming radio data that might indicate the presence of an extraterrestrial signal. So far no obvious alien messages have been discovered, but organizers of the project are hopeful that the enormous amounts of computing power being employed will eventually pay off, perhaps in the next few decades.

Although numerous search programs have been carried out by now, hardly any radio messages have been *sent* from Earth to the stars with the intent of establishing lines of communication. This could explain the lack of messages received: if everyone is listening and no one is sending, then no one should hear anything. One might argue that normal leakage from our planet's radio and television broadcasts will be detected by listening programs; but such emissions are *broadcast*, rather than beamed, toward audiences on the Earth, rather than on distant planets. And there are hundreds of stations broadcasting at the same time, meaning that the aliens will be receiving a *mixture* of programs at each frequency, with the signals fluctuating as the Earth rotates. It is unlikely that any terrestrial leakage will be either strong enough or coherent enough to be detected by extraterrestrials around nearby stars, with radio telescopes similar to ours. In order to be noticed by extraterrestrials, we will probably have to *beam* our signals towards individual stars, concentrating the radiation in a small region of the sky.

Most radio searches have been designed to look for *abnormal* signals – those that are unlikely to be created by natural processes. The primary candidates to meet this criterion are thought to be signals with extremely narrow bandwidths, on the order of 300 Hz or less. Thus, the searches have often looked for *spikes* on the frequency spectrum, beacons that attract the attention of other civilizations but make no real attempt to deliver any information about the senders other than their existence and direction in space. A message with actual content will be somewhat more involved.

TALKING TO ALIENS

Many people erroneously regard radio waves as some special type of *sound* waves (radios do produce sound, after all). They figure that as soon as the aliens point their radio telescopes toward Earth, Elvis will begin crooning over the observatory loudspeakers and Ed Sullivan will appear on all the computer monitors. But it will not be that simple. Successful deciphering of terrestrial radio and television programs by extraterrestrials will require that the electromagnetic radiation that reaches them must be of sufficient strength and clarity to reproduce the original signal *and* that the aliens have proper electronic gear available to generate the appropriate sounds and/or pictures.

The latter point is likely to be a problem. Just as European and American appliances operate on different standard voltages and frequencies, the aliens are likely to have electrical standards and products that are not in harmony with ours. Any assumption that the electromagnetic signals that produce pictures in our televisions will have the same effect in theirs – if they even have televisions – is almost certainly unwarranted.

Even if the extraterrestrials do manage to tune in to Elvis or Ed, they will have a hard time understanding them. Although some sort of vocal communication is to be expected for intelligent aliens immersed in an atmosphere, it is highly improbable that English has become a truly universal language just yet. The aliens may be able to distinguish music from monologue, but they are not likely to get much content from either one. (Try to understand a song in French or Russian or Japanese without having any knowledge of the language.) How can we possibly convey information to beings that do not share a common language with us?

Interestingly enough, there *is* a language that both we and the aliens will understand, and there is a common body of knowledge that we will be able to discuss. Intelligent beings capable of constructing and using radio telescopes will certainly understand a fair amount of science and will be quite fluent in the language of mathematics. They will not use the same mathematical symbols or the same numerals, and may not employ a decimal (base 10) numbering system, but they will be able to do algebra, trigonometry, calculus, and the like. We can count on the aliens to have a knowledge of astronomy, physics, chemistry, and other fundamental sciences that is *at least* as extensive as ours. For example, they should be familiar with the structure of atoms, the chemical elements, the molecular makeup of their own bodies, the properties of waves, the speed of light, the principles of relativity, and so on, and we should be able to exchange information on these topics using our common knowledge as a basis. The trick will be to discuss something of interest while still keeping the message as easy to understand as possible. How would such communications work?

BINARY CODE

Radio transmissions start with a carrier wave of some particular frequency – such as the 1420 MHz line of hydrogen. This wave is then modulated in some fashion to encode the desired information; for example, amplitude modulation is used for the AM radio band while frequency modulation is used for the FM band. The simplest type of code to send is a **binary code**, which assigns one of two values to each bit of information. Various names can be used for these values – 1 or 0, on or off, black or white, present or absent, etc. – the names are not significant. The important thing is that each successive bit has either one value or the other, e.g. either 1 or 0. Information can be conveyed by the *sequence* of values that is carried by the wave.

As an example of a binary code in action, consider the binary number system, which uses only two different digits (0 and 1). In comparison, the decimal number system (with which we are more familiar) uses ten digits (0, 1, 2, 3, 4, 5, 6, 7, 8, and 9). In counting with either system, the same principle is in use. In each place within the number, counting proceeds from the smallest digit (0) to the largest. When the largest digit has been reached, the sequence begins again in that place and the digit in the next place to the left is incremented by one. Table 9.1 illustrates counting from 0 to 17 using decimal and binary systems.

Table 9.1: Decimal and Binary Counting

Decimal	Binary		Decimal	Binary
0	0		9	1001
1	1		10	1010
2	10		11	1011
3	11		12	1100
4	100		13	1101
5	101		14	1110
6	110		15	1111
7	111		16	10000
8	1000		17	10001

The value of each number depends on the digits it contains, the position (or *place*) of each digit, and the value associated with each place. For example, in our decimal system, each place has a value equal to a power of 10: the first place on the right is the ones place, next is the tens place, then the hundreds place, thousands place, and so on. These place values can be written as powers of 10: $1 = 10^0$, $10 = 10^1$, $100 = 10^2$, $1000 = 10^3$, etc. The value of a decimal number such as 5307 is found to be 5 times 1000, plus 3 times 100, plus 0 times 10, plus 7 times 1, or five thousand three hundred seven.

Values of binary numbers are found in a similar fashion, except that the digits are now restricted to either 0 or 1, and the place values are now powers of 2 rather than 10: $2^0 = 1$, $2^1 = 2$, $2^2 = 4$, $2^3 = 8$, $2^4 = 16$, $2^5 = 32$, etc., giving us the ones place, the twos place, the fours place, the eights place, and so on. The value of a binary number such as 11010101 is found as follows:

There are eight places; the value of the left-hand place is $2^7 = 128$. The value of the number is then (beginning from the left) 1 times 128, plus 1 times 64, plus 0 times 32, plus 1 times 16, plus 0 times 8, plus 1 times 4, plus 0 times 2, plus 1 times 1, or 128 + 64 + 16 + 4 + 1 = 213. Thus, the decimal value of the binary number 11010101 is 213; the binary equivalent of the decimal number 213 is 11010101.

Although the binary code produces numbers that are generally longer than their decimal counterparts, and the binary system seems more confusing than our natural decimal system, the binary code is most useful for this application because of its simplified system of values. All the aliens need to recognize is that a given bit can have one of two different values; that simple realization will be sufficient to allow them to receive our messages and possibly even decipher them.

Encoding the Message

We could easily send strings of numbers to the aliens, using the binary code to form binary numbers or just using the code to tap out each number one bit at a time, much as Ralph the Wonder Horse counts by tapping his hoof. In this way we could dazzle the aliens with our knowledge of simple arithmetic. We could even show off a bit and send them sequences of *interesting* numbers, such as the **prime numbers** (2, 3, 5, 7, 11, 13, 17, 19, 23, ...) – numbers evenly divisible only by themselves and 1. This might be better, as prime numbers are not generated by any known astrophysical process and thus would be a good indicator of an intelligent sender.

As entertaining as it might be to send mathematics lessons into space for the extraterrestrials to read, it would be far more interesting if we could tell them something about ourselves. After all, if *we* finally obtain a message that explains that two plus three equals five, we will certainly want to know what sort of beings sent it. Without common words or a means of sending them, communications could be difficult. However, there is another method of communication that transcends any written or spoken language – pictures!

It seems reasonable that intelligent beings that have evolved on a planet around a strong stellar light source will have some well-developed sensory organs equivalent to eyes. And intelligent beings that can design and build radio telescopes will certainly have brains capable of interpreting images produced by their eyes. Thus, we can expect that the extraterrestrials who might receive our messages should at least *consider* the possibility that we are sending them an image.

But how can a picture be sent by radio waves? The waves carry a one-dimensional string of bits, but a picture is two-dimensional. This feat is accomplished every day in a relatively simple manner – in television sets.

The television screen is divided into several hundred horizontal rows of several hundred pixels (picture elements) each. The incoming signal is fed into the back of the picture tube where it generates an electron beam directed toward the screen at the front. Much as a lawn mower cuts the grass in a yard one row at a time, the beam of electrons is swept along each row of the screen in order, where collision of an electron with a pixel causes it to glow momentarily. The whole screen is swept about 25 times per second, giving the sensation of continuous illumination to the eye. In a black-and-white television, each pixel is either black or white, depending on whether or not an electron just hit it.

A black-and-white television screen is an example of a binary code in action. We can use the same principle to send images to the extraterrestrials; a string of ones and zeros can be sent and used to fill in the picture, row by row. Whether the aliens use black for the zeros and white for the ones or vice versa is irrelevant; they will still get the picture (or its negative) – *as long as they know how long the rows are.* And that is the catch.

As noted above, the aliens will probably be working with alien television sets. The number of rows and the number of pixels per row will probably *not* be the same as are found on terrestrial television screens. If only we could tell them the dimensions of the screen, then they could form the picture and decipher the message. But how can we communicate to them the screen dimensions without making that information part of the message?

Remarkably, there is a way to do just that. Consider for example a screen with dimensions of 300 pixels across and 200 rows down. The total number of pixels – the area of the screen – is then 300 times 200 or 60,000 pixels. In order to send an image to fit this screen, we must send 60,000 bits, one for each pixel. The alien receiving such a message could count the bits and thus know the *area* of the screen to be used for the picture *before* actually forming it. Of course, this is not good enough; the correct dimensions will be needed or the picture will not make sense. But the clever alien will note that the area of the screen is the *product* of the two dimensions, and thus would simply need to factor the total number of bits to obtain the dimensions.

In this particular example, the number 60,000 factors *many* different ways: 250 x 240, 300 x 200, 400 x 150, 480 x 125, 500 x 120, 600 x 100, 800 x 75, 1200 x 50, 1500 x 40, etc. While some of these pairs would result in very strange television screens, each of them *could* be used to form an image to send to the extraterrestrials. The aliens might eventually be able to come up with the correct dimensions by a process of trial and error, but in keeping with our intent to make our message easy and obvious for the aliens to decipher, we should probably narrow down the choices a bit.

One option might be to form the picture on a *square* array of pixels, making the number of rows equal to the number of pixels per row. The total number of pixels would then be a perfect square, such as 2500 (= 50 x 50) or 26,569 (= 163 x 163). The aliens would then need only to count the bits, take the square root of the total, and use that value for each dimension. Of course, this plan would not necessarily rule out the other pairs of factors, and the aliens might waste time and effort trying to make them work.

Another plan would be to drastically reduce the number of factoring options by restricting each dimension to a prime number. (See the Appendix for a list of prime numbers.) With this limitation the total number of pixels will be the product of two primes. As prime numbers cannot be factored any further, there will be only one pair of factors for the aliens to choose. For example, the factors of 323 are 17 and 19, both primes. Aliens receiving a message whose total number of bits is the product of two primes will be led directly to the correct dimensions as soon as they count the bits and factor the result.

Messages of this type, in which a picture is encoded as a string of bits, are called **pictograms**. Pictograms have been proposed as a means of sending messages to extraterrestrials and, in at least one case, have actually been transmitted. Before examining any large pictograms, let us consider the following simple illustration.

PICTOGRAM PRINCIPLES

Here is a sequence of ones and zeros, representing the bits in a short pictogram message:

01110011110100010100011000001000110000011110100110100011000101000101110011110

The first step in the process is to count the bits; mathematically astute readers should find 77. The second step is to factor 77, which is easily done; the factors are 7 and 11. The next step is to lay out the bits in a 7 x 11 array to form the pictogram; here is where it gets just a bit tricky. Assuming that the bits are to be inserted into the different rows of the pictogram in the same manner that each line in this book is to be read – left to right, left to right, left to right, etc. – there are two different arrays that can be formed. One has 11 rows of 7 pixels each and the other has 7 rows of 11 pixels each. The two results are as shown in Figure 9.6.

```
      0111001
a     1110100                b
      0101000                   01110011110
      1100000                   10001010001
      1000110                   10000010001
      0000111                   10000011110
      1010011                   10011010001
      0100011                   10001010001
      0001010                   01110011110
      0010111
      0011110
```

Figure 9.6: 7 x 11 array options.

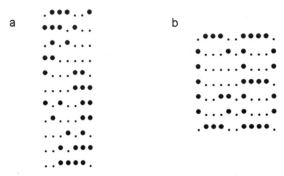

Figure 9.7: Pictogram options.

In Figure 9.6a, the array has 11 rows of 7 pixels each while in Figure 9.6b there are 7 rows of 11 pixels each. The latter array seems to have a bit more order to it, but it is difficult to see patterns with the ones and zeros. This can be remedied by replacing the ones and zeros with different characters, as shown in Figure 9.7.

The visual appearance of each pictogram now makes the correct version much easier to judge. While Figure 9.7a still makes no real sense, Figure 9.7b clearly spells out the initials of a recent American president. Of course, this example is not intended as an actual message to the stars; aliens are not expected to recognize *any* letters of our alphabet although they would probably notice the greater amount of order in the second pictogram.

As noted above, pictograms have been used in the past to illustrate their potential for transmitting information without using language. Following a conference at the National Radio Astronomy Observatory in Green Bank, West Virginia in 1959, Frank Drake devised and sent such a pictogram to all the participants (Breuer 1982). The message, intended as a possible alien communication, was supplied as a string of ones and zeros (as shown in Figure 9.8), with no special instructions as to how to proceed.

```
1110011000001101100011011000001100000000110000010000010001000
1000001000011000110000010000010001000100010000010000000000010000010000 01
0000010000111000010000010000000000010000001000100000101000001010 10
000010111100000000111000001111100000111000000000000010000001000100
0100100100010001000000000000000000100010001010101000100010000010 1
1011011100110110010001000100010001000000000000000000010001010100000 0
00101000100000000000100000010101000000000001010100000001010111000
000011110000000000001111100000000000010000010010000000000000001
00100000000001000010010100000000000000010100111010100110000100101
0000000000000010100000000000011100100010000000000000000010000000
000000000000000000010111110000000000000010011101101000000000010
00001000000000001000000000000000000000101111100000000001000000000
01100001100001100001100001100001001001010100100100100100100100100
10010010000000001000011000011000011000011000011000000000000000000
00000000000000000000000000000001000110000100000100000010000010000000
00000000000000000000000000000000000000000000001000000010000000000
0000100000000000011100000001000100010000000010000001000001000010
0000000001000000000000100000010001000100000000000000000000000000000
000000000010001000000010000010000010000010000010000000000111000010000
00000000000000000000000000000000000000001
```

Figure 9.8: Drake's 1271-bit message.

Interpreting this code as a pictogram begins with counting the bits. There are 1271 of these, a number that factors into the prime numbers 31 and 41. Using these as the dimensions of the array gives two choices: 31 across and 41 down or 41 across and 31 down. Both of these possibilities are shown in Figure 9.9, with the ones and zeros replaced by dots of different sizes.

Once again, it is fairly clear which array is intended; Figure 9.9b looks considerably more orderly, but some might claim that it is upside-down. It only appears this way because the array was filled in by beginning in the upper left corner and working from left to right, as this book is meant to be read. However, there is no guarantee that extraterrestrial civilizations will read their books – if they have books – in the same way. In fact, there are *terrestrial* civilizations that do not use the left-to-right, top-to-bottom convention found in this book. We must be flexible here and willing to consider different orientations in order to see what the aliens are trying to tell us.

Before the pictogram is interpreted, some thought should be given as to what *types* of information it might or might not contain. What information would we want to know about the extraterrestrials? What might we be willing to tell them? What types of information can possibly be transmitted?

Names are probably out of the question; names of planets, stars, alien leaders, etc. will be nearly impossible to convey without language and would not be that helpful anyway (the aliens undoubtedly have different names for the stars we know). Information that is likely to change significantly in a relatively short period of time (such as the population of a planet) will not be very helpful to civilizations receiving the message. Numbers *can* be easily included, but those that are *measurements*, requiring specific units (years, grams, astronomical units, etc.) to be complete, will be much more difficult – but surprisingly, not impossible – to communicate.

The main things we could expect to find in a pictogram will be images, diagrams, and numbers – images of the aliens or other life forms on their planet, diagrams showing the location of their star or the layout of their planetary system, and binary numbers to indicate compositions of aliens, planets, and atmospheres. With these ideas in mind, let us return to Drake's pictogram to deduce its message.

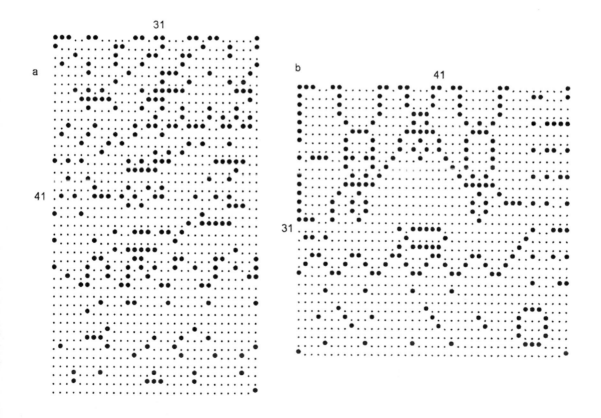

Figure 9.9: Drake's 1271-bit pictogram – array options.

167

For *our* convenience, Figure 9.9b is reproduced in an inverted form, shown in Figure 9.10. Note that the pictogram's *meaning* should not be altered by its orientation with respect to the reader, although the reader's *perception* of the meaning may be.

The most obvious images in the pictogram would appear to be the group of aliens at the bottom – three humanoid figures that we would probably interpret as two parents and a child. The alien on the right seems to be pointing at a dot, one of eight in a column below the circular image in the upper right corner. This group could represent the aliens' star and its planetary system, and the alien could be indicating that they inhabit the fourth planet.

To the right of each planetary dot is a group of dots that can be interpreted as a binary number, with dots as ones and blanks as zeros. At first glance, the binary numbers would appear to read (from top to bottom) 11, 101, 111, 1001, 1011, 1101, 1111, and 10001, corresponding to the decimal numbers 3, 5, 7, 9, 11, 13, 15, and 17. But there seems to be no reason why *these* numbers should be presented and identified with the planets.

There is a problem with binary numbers – actually with *any* numbers – in knowing which digit represents the ones place. *We* normally read numbers from left to right, but the aliens may not do so; additionally, we may not have the pictogram laid out the way the aliens drew it – we could be looking at the mirror image. In order to indicate how to read each binary number, the aliens have included a 'decimal point' (or perhaps, a *binary point*) on each one; the decimal points are lined up along the right side of the pictogram. With this interpretation the binary numbers become 1., 10., 11., 100., 101., 110., 111., and 1000., which are of course the decimal numbers 1 through 8 – reasonable labels for the eight planets in the system.

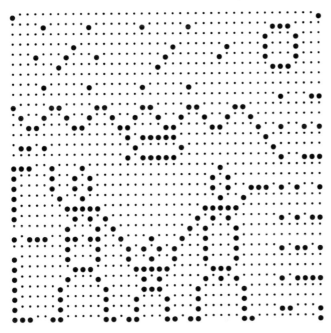

Figure 9.10: Drake's 1271-bit pictogram – correct array, inverted.

Across the middle of the pictogram is a wavy image – perhaps mountains or water waves – that points to the third planet. Below this image is what appears to be either a rocket or a fish of some sort. Linking these two images together, the most appropriate interpretation would seem to be that the third planet has oceans that contain some form of aquatic life (rather than rockets flying under mountains).

Across the top of the pictogram, to the left of the star, are three curious groups of dots that constitute the aliens' chemistry lesson. The first two dots to the left of the star represent the nucleus and single electron of a hydrogen atom. The next group to the left contains a nucleus at the center, two inner electrons, and four outer electrons, as in a carbon atom. The group on the far left is similar, with a nucleus, two inner and six outer electrons, making oxygen. In diagramming hydrogen, carbon, and oxygen atoms, the aliens may be indicating that their biology, like ours, is carbon-based.

To the left of the aliens' images is a bracket that seems to indicate their height. At the center of the bracket is the binary code 10111, oriented in the same manner as those labeling the planets. Reading the last digit as a decimal point yields the binary number 1011., which is equivalent to the decimal number

11. However, a height of 11 is meaningless unless we know the units that apply; what unit of length could we possibly have in common with the extraterrestrials?

Strangely enough, there *is* a unit of length that is known to both parties – the wavelength at which the message was transmitted. Assuming this message was sent at the hydrogen frequency of 1420 MHz, the corresponding wavelength would be 21 centimeters, and the alien height would be 11 x 21 = 231 centimeters, or about seven feet.

Another binary code appears to the left of the 'fish' image, above the alien height bracket. If interpreted as a number, it would have a value of 6, but its proper association is not exactly clear. There would surely be more than six 'fish' on the third planet, and the height bracket already has a number. It could refer to some property of the waves above the 'fish', such as the wavelength or the height of each wave, but it is not obvious why this information would be useful. The only other probable association is with the upraised arm of the nearest alien, in which case it could be telling us there are six digits on the alien's hand.

The only dots not yet discussed are those found in the corners of the pictogram; they most likely serve simply to mark these locations – an indication that the rectangle has been properly set up.

Drake later produced another, smaller pictogram, with a similar message. The 551 bits produce a 29 x 19 rectangle, with the correct layout shown in Figure 9.11. The information contained includes an alien image (bottom center) with a height bracket (lower right), a star (upper left corner) with planets of different sizes below it (left edge), and atoms of carbon and oxygen (top). The binary numbers used here are somewhat trickier because they employ a parity bit – an extra bit added to the number (or not) to make the total number of ones an odd number.

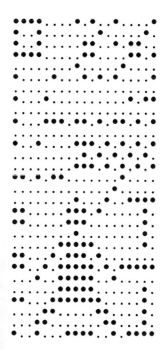

Figure 9.11: Drake's 551-bit pictogram.

These pictograms have been intended as example of messages sent out by extraterrestrials, describing themselves and their planetary systems. On at least one occasion, a pictogram message was actually sent from the Earth, beamed by the Arecibo radio telescope toward the globular star cluster M13 in Hercules. Sent November 16, 1974, this 1679-bit message included a human image, a diagram of the solar system, and empirical formulae for a number of the building-block components of the DNA molecule (Breuer 1982). Any civilizations within this cluster of 300,000 stars will not receive the message for some time, as M13 is 25,000 light-years away. And we need not wait anxiously for a return message, for that will not arrive for at least 50,000 years. Of course, M13 is a Population II object, an old cluster with relatively low metal content, and as such, it is unlikely to harbor terrestrial planets of sufficient size to host intelligent beings. It is also orbiting the center of the Galaxy and thus moving; we see it today where it was 25,000 years ago, and the message was aimed at that position. But it is not there anymore, and it will be even farther away from that position by the time the message arrives 25,000 years in the future.

The 1974 pictogram message has become another message in a bottle – a very fast bottle – with an extremely low probability of being received. Its transmission was mostly symbolic, as it was made during the dedication ceremony that celebrated the conclusion of an upgrade to the 1000-foot radio telescope. M13 was chosen as the target primarily because it lies in a part of the sky that was accessible to the telescope during the time of the ceremony.

Pictograms may prove to be useful methods for communicating with extraterrestrials, but they are not necessarily the ultimate tools. Aliens might choose to devise even more elaborate three-dimensional pictograms, perhaps forming the bit total from the product of *three* prime numbers. Pictograms need not be simple enough for *every* member of a civilization to solve; one individual with a reasonable solution will be sufficient to open a communication channel to a distant alien society. Readers wishing to practice for such an event may attempt to decipher the pictogram message shown in Figure 9.12. A complete solution is given in the Appendix.

```
1110000010101000000011110000010000000100000001100010000000000010
0000100000000000000000000000001000000000000000100000000000000000
0000000000001000001000100000001000000000111000010000000001000000 1
0000000000001000000000000000010000000010000000000000000000100000
0001000001100000000010000000000011110000010000000001000111100000 0
0101000000010011000010110000000000000000000101001000000000000
0010111000110000000000000010010001100000000000100101100000010000
000100001101000010001010100000011110000100000000000000100010000 11
0000000000010011000001000000001100000000000100000100100 100000000110
00000100110010000000000001001111000010000110100111001111100010000
0000011110111101001000011111101111110000110000000001111011110100
1000100010011000111110010000000000000010110000100011110000000100
1110010000000000000100001001001010010000000000000000011000000001111
111111110000000001011
```

Figure 9.12: A sample pictogram message.

INTERSTELLAR CHAT ROOMS?

Interstellar communication with extraterrestrials appears to be possible with the technology we now have, but it is certainly not trivial. There is no guarantee that we will successfully make contact with other civilizations, even if they do exist. If we *should* manage to establish communications with aliens, there still remains the problem of response time. Messages will require decades to get to relatively nearby civilizations; real dialogues may be out of the question. Interstellar chat rooms probably do not exist.

It is also possible that the alien civilization may change over time, perhaps becoming less willing – or possibly unable – to respond to our queries. Although we regard ourselves as a relatively young technical civilization, that may not be the case. How advanced can a technical civilization become? How long can a technical civilization last? These issues will be discussed in Chapter 10.

MAIN IDEAS

- All forms of electromagnetic waves travel through the vacuum of space at the maximum possible speed – the speed of light; for this reason, electromagnetic radiation is considered to be the most efficient means of establishing communications across interstellar distances.

- Additional requirements narrow the choice of a likely communication frequency to the microwave/radio region of the spectrum, where the levels of natural and artificial noise should be sufficiently low that signals could be detected.

- The 'water hole' – a particular range of microwave frequencies bounded by the emission frequencies of hydrogen atoms (H) and hydroxyl radicals (OH) – might be selected as a communication channel by intelligent species that also have a water-based biochemistry.

- Although we do have radio telescopes capable of searching for signals from alien civilizations around relatively nearby stars, most such telescopes are currently used for regular astrophysical research; hardly any telescope time is devoted to SETI programs.

- While extraterrestrial technical civilizations are unlikely to understand any of the languages on Earth, they should still have sufficient knowledge of mathematics and science to allow the sharing of some information with us, if they choose to do so.

- Pictogram messages provide an example of the manner in which information might be transmitted to (or by) an alien technical civilization; their use requires little more than suitable radio telescopes, recognition of binary code, and some imagination.

KEYWORDS

bandwidth	binary code	Doppler shift
electromagnetic radiation	electromagnetic spectrum	energy
frequency	noise temperature	photon
pictogram	prime numbers	Project Ozma
SERENDIP	SETI	speed
water hole	wavelength	

LAUNCHPADS

1. Suppose that radio waves and microwaves could not penetrate Earth's atmosphere. What effect would this have on our search for extraterrestrials?

2. Would it be irresponsible for us to send radio messages to the stars, announcing our presence to friend and foe alike?

3. Suppose an extraterrestrial civilization exists on a planet orbiting Alpha Centauri. Are we likely to become radio pen pals with them?

REFERENCES

Breuer, Reinhard A. 1982. *Contact with the stars.* San Francisco: W.H. Freeman.

Chapter 10

THE RISE AND DEMISE OF TECHNICAL CIVILIZATIONS

*I*n which possible levels of technology are described, biological and planetary prerequisites for technical civilizations are explored, a variety of limiting factors for technical civilizations are presented, and anticipated lifetimes for technical civilizations are discussed.

The objects of our search are extraterrestrial *technical* civilizations, defined (in this book) as those civilizations that are capable of interstellar radio communications. In using this particular definition of 'technical', we have managed to include ourselves in this category; however, it is not certain that extraterrestrials will use the same criterion to select civilizations of interest to them. In their opinion, we may still be unworthy of contact, even though we are perfectly capable of communicating with them using our radio telescopes. Civilizations with technical capabilities far in advance of our own could have developed in our Galaxy; perhaps their existence has produced clues for us to find, allowing us to detect their presence without actually making contact with them.

LEVELS OF TECHNOLOGY

With a history of radio astronomy research spanning only a few decades, our civilization is just barely technical in terms of its ability to send and receive interstellar messages. Other measures of technology could also be considered, such as space exploration capabilities or energy consumption, but this information about the extraterrestrials will not be so easily ascertained. Even so, we could still speculate about such characteristics of alien civilizations, with an eye to particular levels to be attained.

Such speculations were made in 1964 by the Soviet astronomer Nikolai Kardashev, who used the total energy consumption of a civilization as the defining criterion for each technology category. One *could* set these criteria at some rather arbitrarily defined levels, but Kardashev elected to match the energy levels to particular energy *sources*, producing what are known as **Kardashev civilizations** (Breuer 1982).

The primary energy source for a planetary civilization will usually be the star about which the planet revolves. Radiant energy from the star falls on the planet, where it is absorbed by the atmosphere, the surface, and various forms of life. Technical civilizations may utilize this energy in a variety of ways: by consuming plants (and the animals that feed on them); by generating hydroelectric power from water flows resulting from the planet's water cycle; by building collecting panels to turn solar/stellar radiation into electricity, etc. They may also supplement this source with nuclear energy, geothermal energy, wind power, and so on. Civilizations that utilize energy at about the same rate as our own are classed as Kardashev Type I civilizations. This value — set by Kardashev at about 10^{12} watts — is fairly flexible but quite representative of the Earth's current energy usage.

Other Type I civilizations could be expected to be engaged in activities similar to those of our terrestrial civilization. They too should be capable of interstellar radio communication and some space travel within their planetary systems, but they would most likely not be voyaging to the stars.

A civilization seeking significantly more energy for extremely ambitious projects might attempt to collect a much greater fraction of the radiation emitted by its parent star. The Earth intercepts only about 1/20,000,000,000,000 of the Sun's radiation, leaving *considerable* energy available for expansion beyond the Type I level. Of course, actually *obtaining* this energy will require engineering on a much grander scale than any Type I civilization can manage; consequently, Kardashev set the criterion for Type II civilizations as the utilization of the power output of an entire star, about 10^{26} watts.

In order to collect all of its star's radiation, a Type II civilization would have to construct a huge sphere around the star, intercepting every beam of starlight, to power current projects or to be stored for future use. Such spheres are termed **Dyson spheres**, after the physicist Freeman Dyson who proposed their existence. The mass needed for such a sphere would be substantial; for example, a shell with a radius of one AU and a thickness of one meter would require a volume of matter equivalent to that of a spherical body with a diameter of 81,000 kilometers (midway between Saturn and Uranus in size – about 260 times the volume of Earth). Reducing the thickness to only one centimeter would cut the volume requirements down to 2.6 times that of Earth – still a significant amount of matter to be mined, processed, fabricated, transported, and assembled into a functioning Dyson sphere. Presumably the Kardashev Type II civilizations will be up to this challenge, having spent considerable time and effort developing their space engineering techniques.

Such a project could not be done overnight; most likely there would be construction stages lasting for decades or even centuries. The civilization would need a plentiful supply of skilled workers eager and willing to spend significant fractions of their lives in weightless or near-weightless conditions. Remote bases for the workers would need to be established on other planets, moons, or asteroids, or in artificial space colonies. Huge fleets of space vehicles would have to be produced and supplied for the construction effort; much of the energy gleaned by initial phases of the Dyson sphere would go toward fueling the construction fleet. The normal activities of the planetary civilization would gradually become a very minor component of the total economic picture of the Type II civilization.

It would seem that a civilization attempting the leap from Type I to Type II would need an extremely strong motivating factor to keep its citizens focused on such an immense project, undertaken not for their benefit and lasting much longer than their own lifetimes. The ultimate goal of the Type II civilizations may well be to obtain sufficient energy to power their interstellar missions – another long-term project of questionable value for the citizenry left back home. It is difficult to conceive of the success of such efforts, given the politics involved in obtaining funding for scientific ventures on *this* planet; but then, perhaps Type II civilizations will be totalitarian rather than democratic governments, capable of accomplishing great feats without obtaining the consent of the populace.

If Type II civilizations do exist, we may be able to detect them. A circumstellar sphere that collects all of its star's radiation will gradually heat up until its own thermal emission balances that received from the star, less the energy tapped for use by the Type II civilization. In equilibrium, a Dyson sphere could have a luminosity as high as that of the star it encloses; however, with its much larger radius, the sphere would have a considerably lower temperature. For example, a sphere around the Sun at a radius of one AU would have an equilibrium temperature a few degrees above the boiling point of water, far below the Sun's normal surface temperature of several thousand degrees. At this lowered temperature the Sun's Dyson sphere would emit most of its energy in the infrared portion of the spectrum and thus appear to outside observers as an infrared object, rather than a visible star. Of course, natural infrared sources also abound, complicating the process of detection, but there is some hope of identifying such civilizations – if they exist – without relying on the interception of radio signals. To date, none have been found.

A civilization that outgrows its Dyson sphere will have quite a ways to go before reaching the next level of technology; Kardashev's Type III civilizations are those that can consume the entire power output of a galaxy – roughly 10^{37} watts. While the efforts of Type II civilizations would be largely confined to their own planetary systems – moving, dismantling, and reconstructing planets, asteroids, and the like, perhaps with a goal of fueling and provisioning their interstellar voyages – the Type III civilizations would be expert space travelers capable of moving throughout their galaxy and controlling every aspect of its operations. These civilizations would have existed long enough to have spread throughout their home galaxy, probably cataloging and monitoring the other civilizations they encounter.

The power available to Type III civilizations would permit them to engage in activities on such a grand scale as to appear incomprehensible to Type I civilizations such as ours. Indications of a Type III presence in our Galaxy might be provided by observations of abnormal phenomena that, despite our best efforts, resist explanation by any natural processes. A Type III civilization that takes an active part in the affairs of a lesser civilization might appear godlike to its citizens. Presumably, such extraterrestrials would be perfectly capable of remaining undetected by us if they so desired; of course, with the power they would wield, it is not clear why they would worry about being noticed by lesser civilizations.

Defining levels of technology is one thing; attaining them is another issue entirely. While we know that Type I is certainly within reach (as we are already there), it is not at all clear that Types II or III are even feasible, given what we know of the difficulties of space travel. As creatures of the Earth, humans have been able to adapt to a variety of terrestrial environments, often by employing various technological tools to enhance our chances of survival. We have also managed – using even more technology – to adapt to short-term survival in space, but it is not obvious that *long-term* survival in space can be satisfactorily achieved by bodies originally optimized for existence on the Earth's surface. Although it may be unwise to declare Kardashev's advanced levels of technology to be impossible, it may be equally unwise to presume that they can be achieved by *any* civilization that outgrows its home planet.

REQUIREMENTS FOR TECHNICAL STATUS

The development of an extraterrestrial technical civilization places some requirements on the structure of the alien bodies and others on the aliens' environment. Failure to meet these conditions will probably prevent technical civilizations from arising even though intelligent life may in fact be present on a planet.

Biological Standards

It is tempting to declare that aliens who successfully develop technical civilizations will be likely to have bodies similar to ours, and the rash of humanoid aliens in the movies has done little to squelch this idea. There are certainly some advantages to our body plan – our relatively large brains and versatile vocal cords allow us to communicate with a variety of languages; our upright stance places our head with its sensory organs at maximum height above the ground and frees our arms and hands for manipulating tools; our opposable thumbs provide an improved means of gripping tools and other materials; our oxygen-breathing lungs permit our existence in the Earth's atmospheric blanket, outside of the oceans where terrestrial life developed; and so on – but surely there must be other ways to package the same features outside of a humanoid form.

Other reasonably intelligent species have evolved on the Earth, but they have lacked one or more of the key characteristics of humans that allow our technical status. For example, dolphins appear to be fairly intelligent, but they have flippers instead of hands, and they live in the water where fire, electricity, and radios are all extremely difficult to invent. There are many ways in which the biology of a species can thwart its development of technology, but there is probably no way for biology to guarantee that technology *will* arise in a given line of evolution. As with intelligence or any other desirable biological feature, technology may have to wait for its chance in the evolutionary lottery. The odds of developing a potentially technical species may be very small, but the evolutionary dice are continually being rolled. With enough chances, it can be argued that such a species will eventually appear on a planet. Whether 'eventually' will be soon enough is one issue; whether the planet will actually cooperate is another.

Planetary Standards

Evolution of an optimal body type together with adequate intelligence may not be sufficient to insure development of a technical civilization, as there are a number of critical features that the planet must contribute beyond simply being suitable for life. It has already been noted that intelligent life that is confined to the water will not be likely to develop the necessary skills and equipment to become technical; therefore, a planet that is completely covered with oceans will be an improbable site for a technical civilization. Earth's ratio of water/ice to dry land is currently about 78% to 22%, and this has clearly been satisfactory for us; this ratio has undoubtedly varied over the eons as ice ages have come and gone and Earth's climate has changed, but the variation has not been dramatic over the last few hundred million years, during which our current complex life forms have developed. Just what the operational limits are is unclear; whether 10% (or 1% or 0.1%) dry land would provide sufficient area for a technical civilization to develop is quite unknown. On the other end of the scale, we do not know the minimum amount of *ocean* necessary for the evolution of life on a planet, but there certainly must be one. The limits on this range of ocean fractions will play a key role in determining the value of the factor f_i in the Drake equation.

There were intelligent species on Earth long before our technical civilization evolved; stone-wielding hominids apparently existed a few million years ago, but the transformation from the stone age to the bronze age – when humans began using metals to fashion superior tools and weapons – occurred only about 6000 years ago. The leap to our current technical age required a number of different metals, such as copper (the key element in bronze), which is used in virtually every current electronic device. Ores of copper and other metals are typically found in mountainous regions of the Earth as opposed to prairies, river valleys, coastal regions, etc.; mountains are created by the process of plate tectonics, which causes the drifting continental plates to collide, pushing up ranges such as the Rockies and the Andes. Were it not for this mountain-building process, the deposits of metallic ores would lie so deep beneath the surface that humans might never have discovered them. Living on a planet without plate tectonics, an intelligent civilization may be condemned to the Stone Age, no matter how clever its citizens may become.

Another major factor that supported our terrestrial civilization's drive toward technology was the discovery of fossil fuels. Deposits of coal, oil, and natural gas have provided improved fuels for the development of industry and technology. Even more important, petroleum derivatives have been the principal raw materials of the plastics industry, which has been essential to the manufacture of components used in computers and other electronic devices, satellites and space probes, and a large proportion of the products in our modern technical world. Given the major role fossil fuels have played, it is difficult to conceive of our civilization's achieving its current level of technology without any access to these underground stockpiles of organic material; will extraterrestrial civilizations be so fortunate?

Our fossil fuels probably originated a few hundred million years ago when forests of early plant life flourished on the Earth's continental landmasses. The deaths of these plants, the submergence of their remains in shallow swamps (preventing normal oxidation processes), and their subsequent burial by accumulating sediments produced the high pressure conditions that ultimately transformed these collections of organic debris into today's deposits of coal and petroleum. Thus, our mode of transportation to the mall is highly dependent on the fortuitous fate of primitive plants that lived a few hundred million years ago.

Presumably, extraterrestrial civilizations will also have evolved over a lengthy period of time, and presumably they will have been preceded by plant species similar to those that ultimately formed our fossil fuels. But it is *not* obvious that geologic and climatic circumstances on the aliens' planets will combine to form similar deposits of fossil fuels, nor is it clear that such deposits, if formed, will necessarily be readily accessible to the alien civilizations. As in the case of the metal ores, lack of fossil fuels could turn out to be a major impediment to a civilization's drive toward technology. These issues, together with the biological factors should be incorporated into the estimate of the value of the intelligence/technology factor (f_i) in the Drake equation.

LIMITING FACTORS ON TECHNOLOGY

It is most likely that not every intelligent extraterrestrial species will develop a technical civilization; those that do will face a variety of different challenges as they attempt to *maintain* both their technical status and their existence as intelligent species. In examining these challenges, we will consider our own situation here on Earth, examining the key elements of our own technical existence and watching for potential hazards on our horizon.

Astronomical Accidents

There are a number of astronomical catastrophes that may befall our planet, with potentially deadly consequences for humanity. These are the types of disasters that a Type I civilization such as ourselves will be ill-equipped to prevent, for they generally involve either some inevitable stage of stellar evolution or the normal orbital motions of the myriad of celestial bodies.

We have seen in Chapter 6 how stars evolve at rates dependent on their masses; stars with the Sun's mass spend about 10 billion years on the main sequence before they expand to become giants. With the solar system's current age of 4.6 billion years, we have about 5 billion years to go before the Sun swells to its giant phase and broils whatever life may still exist on Earth by then. More likely, terrestrial life will begin to feel the heat from the Sun's gradually increasing luminosity much sooner than that – perhaps in only a billion years or so – as the inner edge of the continuously habitable zone moves out to 1 AU. This problem should be fairly well understood and quite predictable for any technical civilization facing the demise of its star. Whether the aliens will be able to do anything to escape this peril will probably depend on the level of technology they have attained by then. In most cases, this should be a long-term problem and therefore not a major concern for technical civilizations (unless they have foolishly developed in the planetary system of a more rapidly evolving F star).

Although most stars do not pose any real threat to our civilization, the most massive stars certainly may. When these stars die, they explode as supernovae, sending particles and radiation streaming into space. Such emissions from nearby supernovae may have caused mutations that altered the course of evolution on Earth; intense radiation reaching our technical civilization could fry our communications networks, destroy our ozone layer, and have adverse biological effects as well. Fortunately, supernovae are relatively infrequent in our Galaxy – estimated rates are about one or two per century – with *nearby* supernovae even less likely. But over a long span of time such explosions could occasionally cause problems for civilizations on planets of neighboring stars. Our civilization has not yet developed a good way to predict impending supernovae other than to identify likely candidate stars; perhaps astronomers will eventually come up with a good supernova indicator that will provide an exposed civilization with plenty of warning and time to prepare for the effects of such an event. Of course, a civilization on a planet so close to a supernova as to be immersed in its supernova remnant will probably not fare well, even with advanced notice.

Of a similar nature are the gamma ray bursts, noted in Chapter 6. Although extremely rare, these events are so energetic that the consequences for advanced life anywhere in the same galaxy when one occurs could be quite dire. Some have proposed such a burst in our Galaxy as the cause of the extinction of the dinosaurs about 65 million years ago, but there is no conclusive evidence to support this hypothesis. It may be that at least some of these bursts happened *only* in the early stages of galaxy formation, meaning that they are unlikely to occur at present in the Milky Way although we can still detect them in very distant galaxies. Just when the next nearby gamma ray burst may occur and the effect it will have on existing technical civilizations are both completely unknown at this time.

The Sun and all the other stars in the Milky Way are orbiting the center of the Galaxy on time-scales of a few hundred million years. The Sun's essentially circular orbit in the plane of the Galaxy's disk is modified by the gravitational field of the nearby disk stars, producing a small oscillation above and below the Galactic plane. This complex motion of the Sun and other stars creates the potential for stellar orbits to intersect the numerous gas and dust clouds that populate the Milky Way's disk. Such an interaction could trigger a nonviolent extinction of the unlucky civilization whose planetary system drifts into a dense interstellar dust cloud. The resultant dimming of the planet's sunlight could initiate climatic changes that would dramatically alter the planetary environment, making it totally unsuitable for the technical civilization (see further discussion below).

Interactions between neighboring galaxies occur on time-scales of a few hundred million years. Such galactic collisions must certainly have some major effects on planetary systems in the region of the interaction, even though collisions of stars and planets are highly unlikely. Gravitational tidal forces may disrupt the orbits of planets, asteroids, and comets, increasing the impact rate on a civilization's home planet and possibly terminating their technical status. In fact, even without such large-scale interactions, impacting bodies pose a major threat to planetary life.

Formation of our planetary system involved the process of accretion, in which small bodies were swept up by collisions with larger bodies. These encounters produced impact craters, such as those easily visible on the surfaces of Mercury and the Moon. Most of the impact craters on Earth have long since been erased by the action of plate tectonics, which gradually and relentlessly rejuvenates the surface. However, a few dozen impact features have been identified on various continents, most of them so severely weathered by wind and water that the process that created them has not always been obvious. These features are comparatively recent additions to the Earth's surface, with the youngest one – the Barringer meteorite crater in Arizona – believed to be about 50,000 years old.

The Earth is constantly being bombarded by **meteoroids**, most of which are tiny rock fragments that burn up in their passage through the atmosphere, creating the **meteors** (or shooting stars) we see in the night sky. Somewhat larger meteoroids land intact on the surface as **meteorites**, to be discovered some time in the future by curious humans. Only a few very large meteorites have been found because (1) collisions with large meteoroids are rare, due to their lower abundance in the solar system, and (2) very large meteoroids are more likely to be broken up or vaporized by the greater energies involved in their impacts. However, it is becoming fairly clear that the Earth still collides with the occasional meteoroid that is big enough to do serious damage to terrestrial life forms. The source of these large meteoroids is the swarm of asteroids that orbit the Sun. Our further discussion will refer to collisions with such objects as **asteroid impacts**.

Unlike floods, earthquakes, volcanic eruptions, and similar disasters, asteroid impacts are apparently quite rare, occurring at typical rates of one every hundred thousand years or more (depending on the size of the asteroid). As a result, human culture has no memory of these events, and we have had to discover this peril anew. If an asteroid had landed on a major city once every century, there would be considerably more public awareness and concern, along with plenty of historical literature and legends to study. Instead, we must make do with our small collection of terrestrial impact craters and our ever-increasing knowledge of asteroids as we try to estimate when the next big rock might strike us.

The problems with asteroid impacts are several. First, the crater formed is generally much larger than the asteroid itself. For example, the Barringer meteorite crater, measuring about 1200 meters across and 170 meters deep, is estimated to have been caused by the impact of an asteroid only 25 meters in diameter. Impacts by asteroids several *kilometers* across would be devastating. Second, the effects of an asteroid impact are likely to be felt around the globe, depending on the location of the impact. An asteroid landing in the ocean would trigger **tsunamis** – often called *tidal waves* – that could destroy coastal cities on the shores of that ocean. An impact on a continent would eject debris into the atmosphere where it could dim the Sun's light for months or years, cooling the planet and killing off plant life. The heat and blast wave generated by the impact could ignite forests and cities and level any structures in the vicinity of the event.

Clearly, asteroid impacts have the potential for the destruction of our technical civilization, and perhaps our species as well. There is considerable evidence that an asteroid impact 65 million years ago terminated the dinosaurs' reign on this planet and paved the way for the rise of mammals and the eventual evolution of humans. The site of this impact is identified as a crater with a diameter of at least 170 kilometers at Chicxulub, Mexico, which could have been formed by an asteroid only 10 kilometers across.

The impact that destroyed the dinosaurs also marked the elimination of about 70% of all species on the Earth at that time; extinction events of this magnitude are quite rare, estimated to occur every 100 million years or so. There is some conjecture that the earlier mass extinctions that are indicated by the fossil record might also have been caused by asteroid impacts. It may well be that we humans owe our existence on this planet to these occasional large impacts, which serve as species erasers, wiping the slate nearly clean and allowing the remaining species to expand into the vacated ecological niches. The primary issue for us at this time is when the next big impact might come along and possibly erase *us* – or at least our technical civilization – from the planet.

If all of the asteroids in the solar system could be identified, and if all of their orbits could be determined with a sufficiently high degree of accuracy, then we should be able to predict their collisions with Earth. Unfortunately neither of these conditions have been met. The several thousand asteroids discovered to date probably include the largest ones in the inner solar system, but there are undoubtedly many more that are too small to be easily detected and yet plenty large enough to pose a threat to our existence.

As it now stands, we do not know whether the next big impact will occur in 10 years, 10 thousand years, or 10 million years. Nor is it clear what we could do about it if we had such knowledge. Explosions powerful enough to deflect the course of an asteroid are currently prohibited by international treaties. And a detonation *inside* an asteroid that would break it into a host of smaller asteroids would be the equivalent of turning a rifle bullet into a shotgun shell; at close range, the effects would be different, but probably equally devastating. While a Kardashev Type I civilization such as ours appears ill prepared to deal with an incoming asteroid, a Type II civilization would have no problems at all. The question will be whether a Type I can last long enough to become a Type II.

Comets also populate the solar system, in numbers comparable to those of the asteroids. They too can create havoc by crashing into the Earth, but their mode of destruction is apt to be somewhat different. Being composed primarily of volatile materials kept frozen by the cold of outer space, comets are more likely to *explode* upon entering our atmosphere, without forming an impact crater or leaving solid meteorites to be studied. Such an event occurred in the Siberian wilderness in 1908 when a comet (or possibly an asteroid – the scientific debate has not yet been settled) detonated several hundred meters above the surface, flattening forest trees but leaving no crater to mark the site. The remote location of the **Tunguska event**, as it is known, kept its damage to a minimum. Should a similar collision occur today over a more populated region, the effects could be considerably more serious. An encounter with a stream of cometary fragments – as Jupiter had in 1994 – could prove even more disastrous to a technical civilization.

Impacts by asteroids (and perhaps comets) appear to have played a key role in our existence, destroying old species while triggering the evolution of new ones. In other planetary systems where these bodies exist – and there is no reason to believe that our solar system has a monopoly on such objects – they may well play a similar role. (It could be that our asteroids and comets owe their presence to Jupiter, which prevented asteroids in the belt from forming into a planet and probably flung most of the comets into their current locations in the Oort cloud. But if planetary systems with Jupiters are fairly common, then this is not a problem.) This interaction between planets and asteroids/comets may be crucial to the development of intelligent extraterrestrial civilizations. At the same time, these interactions may also serve to limit the tenure of technical civilizations on planets throughout the Galaxy by rendering planets uninhabitable by advanced beings for lengthy periods of time.

The asteroid impact rate may prove to be the determining factor for the existence of technical civilizations on qualified planets. A rate too low might not give evolution enough chances to start over, producing a planet dominated by species such as dinosaurs that have little hope of developing an intelligent civilization. A rate too high may not give a suitable intelligent species the time it needs to build a technical civilization, or it might snuff out such civilizations soon after they evolve. The impact rate on a life-bearing planet will depend on the sizes, locations, and orbital parameters of the various planets and asteroids in a given planetary system and thus will be rather difficult to determine. It would probably be foolish to presume that other planetary systems will be endowed with the same configuration that has resulted in our development on this planet; indeed, we have evidence to the contrary in the extrasolar planetary systems discovered to date.

Perilous Planets

As we are occasionally reminded, our planet is a dangerous place to live. The Earth is geologically active, with tectonic plates cruising slowly across its surface, driven, in part, by sea floor spreading at the mid-oceanic ridges. These ridges lie directly above hot spots in the mantle, as do some of the active volcanoes found scattered around the world. Other volcanoes are produced by **subduction zones**, where plates of denser rock are pulled (by gravity) beneath plates of less dense material, recycling the denser crustal rock back into the mantle. Individual volcanic activity of the type we normally experience is probably not sufficient to destroy a technical civilization although those regions near erupting volcanoes could certainly be adversely affected. Volcanic ash, injected into the atmosphere and spread by high-level winds, could block sunlight and hinder plant growth on a global scale if enough violent eruptions occur simultaneously. However as active volcanic regions are not particularly difficult to detect, a technical civilization can probably survive by avoiding such danger zones, or at least not concentrating in them.

Supervolcanoes are another issue entirely. These are created where molten rock rises from the mantle into the crust and collects there, rather than erupting immediately as in a normal volcano.

Geologists have found evidence of supervolcanic eruptions in the distant past, the most recent being the Toba eruption in Sumatra, about 74,000 years ago. The scale of this event was unlike anything in human memory; it is estimated to have been 10,000 times as powerful as the 1980 eruption of Mt. St. Helens. Such an explosion could have had far-reaching consequences for life on Earth — not as serious as an asteroid impact, but quite disastrous nonetheless.

There is even some evidence that such effects did occur. Studies of the similarities in mitochondrial DNA indicate that our current population all evolved from a small number of humans about 70,000 to 80,000 years ago — the human population bottleneck. What could have reduced our ancestors to such a small pool during this particular interval of time? The Toba supervolcano eruption is a prime candidate.

The obvious question is when and where the next supervolcano might erupt. One of the largest supervolcanoes known lies underneath the Yellowstone Park region in North America, where it has erupted periodically at intervals of about 600,000 years. As its last event was 640,000 years ago, the next eruption could come at any time, with particularly grim consequences for the United States. Whether our technical civilization will find some way to predict these eruptions or to minimize their impact remains to be seen. If we do not, our civilization may be in for a terrible shock and a short ride.

Earthquakes are another byproduct of plate tectonics. As the various plates attempt to slide past each other, they get stuck because their edges are not smooth. As time goes on, the tension at these plate boundaries builds up until a sudden slippage occurs, causing an earthquake. The longer the time between slippages, the greater the tension and the more violent the resulting tremors will be. Earthquakes can destroy buildings, bridges, dams, and other structures and can cause interruptions in nearly every phase of life for civilizations that lie along fault lines. Quakes of the magnitudes to which we are accustomed are unlikely to bring down a technical civilization or pose a major threat to the human species, but they can certainly make life a riskier proposition. Perhaps advances in research and improvements in technology will make accurate predictions of quakes a reality, and they will become less of a threat to the population. On the other hand, we may discover evidence of super earthquakes — events that could be far more violent than any quakes in human memory, perhaps triggered by asteroid impacts — which could go a long way toward destroying a technical civilization.

Super earthquakes are speculative at this time; there is no record of such events occurring. However, there is ample evidence for another type of global threat: changes in the Earth's climate such as global warming and ice ages.

At the time of the dinosaurs, the Earth's climate was considerably warmer than it is today. For the last 50 million years, the Earth has been getting cooler, forming the Antarctic ice cap about 15 million years ago and northern hemisphere ice sheets about 2 million years ago. Since then the northern glaciers have advanced and retreated four times, covering significant portions of the northern temperate regions with ice sheets up to two miles thick during each of these ice ages. The most recent glacial period began about 100,000 years ago and lasted until 11,000 years ago; the interglacial period that followed has marked the rise of modern human civilization.

Another ice age may be coming — perhaps in a few dozen millennia — but no one really knows. Ice age prediction has no track record as yet. If one *should* occur, it will happen fairly gradually, possibly giving our technical civilization ample time to plan a strategy for survival. The ice age will bring another advance of glaciers, this time into heavily populated areas of the planet — literally wiping them off the face of the Earth; the climate will become much colder, shifting the planet's habitable regions toward the equator; and the sea level will drop by several hundred feet as more of the Earth's water supply is deposited in the continental ice sheets. All of these changes will combine to force considerable modifications to the life style of every terrestrial civilization still in existence. A technical civilization may use most of its resources preparing for the relocation of its cities and the reestablishment of its food supply, which will be necessitated by the advancing glaciers; or it may ignore the peril for centuries until it is too late to take effective action.

On the other hand, we may be in for the opposite effect. The burning of fossil fuels by our current civilization appears to be increasing the levels of carbon dioxide in our atmosphere and thus enhancing the greenhouse effect. If this trend continues, the resulting global warming may melt our polar ice caps over the next few decades and raise sea levels around the globe, again with disastrous consequences for our technical civilization. Global warming also occurs naturally, at the end of each ice age; the exact mechanism that causes this climatic cycle is not yet well understood.

Both global warming and ice ages provide distinct challenges, not only for intelligent civilizations but also for *all* life forms on a planet. As any technical civilization must ultimately be linked through the food chain to lower forms of life, its survival of these disasters will depend heavily on the survival of a sufficient fraction of the planet's biological system.

Harmful Humans

The above discussion has involved hazards that are primarily independent of the existence of intelligent civilizations – they may or may not occur, whether we exist or not. This next group of problems consist of those *caused by* humans; if our civilization is terminated by one of these, we will have only ourselves to blame.

Earlier in this chapter it was noted that our civilization has made heavy use of fossil fuels. We have been mining coal for a few centuries and pumping petroleum out of the ground since the mid 1800s. Our demand for petroleum products in particular, as fuels and as the raw materials for plastics and other materials, has skyrocketed over time, with no end in sight. And that is the problem. The Earth's supply of fossil fuels is not unlimited, and accessible deposits will not last forever.

How long will it be before our civilization has consumed all of the oil on the planet, converting complex hydrocarbons into carbon dioxide and water? Will our technical civilization be able to continue after the wells run dry and the mines are exhausted in a few more decades, centuries, or millennia? Will the public agree to conservation efforts aimed at preserving fossil fuels for the generations to come? There are not many other options. Fossil fuels can hardly be considered renewable resources, as they require hundreds of millions of years to produce. And we are not likely to discover oil deposits on the Moon or any of the other lifeless bodies in our solar system. We can only hope that other sources will be developed to replace our gradually diminishing supply of fossil fuels.

We could turn back the clock and rely on renewable resources again – wood, whale oil, and such – but our current economy is geared to using these resources much faster than they can be generated. We cannot really afford to obliterate whole species of plant and animal life for our own short-term economic benefit – unless, of course, short-term economic benefit is our ultimate goal.

Another dwindling resource that is not given much thought by the general public is topsoil. This is the very thin layer of fertile soil in which most of our crops are grown. Those parts of the world that contain adequate topsoil (and suitable growing conditions) are the principal grain-producing regions for our civilization, and they are obviously quite vital to our existence.

One of these regions – the American Midwest – was covered with an average of 12 to 16 inches of topsoil when settlers began farming there in the 19th century. However, due to the erosive effects of wind and water, this layer has shrunk to less than half its original depth in the heavily farmed areas. At the current rate, the supply of topsoil in the nation's breadbasket will be depleted in only a few centuries, if not decades, producing a major crisis in the food production industry. Expanded attempts to grow crops on marginal soil will increase the demand for inorganic fertilizers, which may ultimately become scarce as well.

Even if conservation efforts are accelerated to slow the erosion, even if topsoil is some day dredged from the delta of the Mississippi River and recycled back north to be spread on the farmland of the Midwest, this may turn out to be a problem that will consume a large portion of the energy and resources of a technical civilization as it struggles just to keep its citizens fed. Funding of searches for alien civilizations may become an unaffordable luxury as natural resources are relentlessly depleted.

As long as humans were still thinking their way to the top of the terrestrial biological system – and there are probably some who would argue that we are not there yet – our numbers were insufficient to do much serious damage to the planet and the other species on it. However, with success has come a much larger population and with it, the potential to have a strong impact on our environment and the life forms that share it with us. One of the most far-reaching of these potentials has been our ability to pollute.

While human numbers were relatively low, the environment had no trouble dealing with us. A few burning campfires or piles of discarded animal bones did not pose any real global threat. However, the coming of modern industry and the concentrated multitudes of people needed to support it have placed a tremendous strain on the environment's ability to absorb the various forms of pollution we create while still maintaining satisfactory living conditions for all life on the planet.

Solid and liquid industrial, municipal, and agricultural wastes dumped into rivers, lakes, and oceans threaten the health and existence of aquatic organisms within these bodies of water, the land-based and airborne predators that feed on them, and other species that depend on these bodies for water (including humans). If we are not careful, our technical civilization may gradually poison our normal water supply, resulting in our eventual downfall (and perhaps the rise of species that thrive on our pollutants).

Similarly, the gaseous waste products of modern industry and consumerism have injected a host of different 'foreign' molecules into our planet's atmosphere. While many of these simply contribute foul odors to our air, others are detrimental to our protective atmospheric blanket, attacking and breaking apart the molecules that form the Earth's **ozone layer**. The ozone gas (O_3) that forms this layer conveniently absorbs the harmful ultraviolet rays from the Sun; reduction of this layer would permit more of these rays to reach the Earth's surface, where their absorption by human skin would cause increased cases of sunburn and skin cancer.

For many years, the chlorofluorocarbon compounds (CFCs) used as propellants in aerosol cans and refrigerants in air conditioners were routinely released into the atmosphere and allowed to rise to the ozone layer to react with its protective, but unstable, molecules. Because of the enormous volume of our atmosphere and the time lag involved, the results of this global experiment were not immediately apparent. However, evidence now exists for the formation of an 'ozone hole' over the southern Polar Regions and reduced levels of ozone elsewhere. Should this hole enlarge to cover a significant portion of the populated regions of Earth, life here could become considerably less comfortable, if not unbearable.

Even though production and consumption of CFCs and other ozone-depleting chemicals have been restricted by treaty since 1989, the effects of the molecules currently in the atmosphere will be felt for some time to come. Of course, this may not be a problem if humans can manage to evolve a more protective skin in the time remaining before the ozone layer becomes completely ineffective.

Although we have finally come to acknowledge and address the problems posed by pollution, we have been slow to do so. Unfortunately, there are many developing nations where the pressure to build local industry to raise the level of technology is stronger than the commitment to keep pollution in check. It may be that a successful technical civilization will require global control of all inhabitants and corporations in order to keep from being undermined by its own pollution problems.

Personal Problems

One of the goals of living beings is to *reproduce*, in order to insure the survival of the species. Organisms at the bottom of the food chain tend to produce many more offspring than can possibly survive because most of them are destined to be eaten before they can reproduce. Those species less likely to serve as prey can afford to concentrate their reproductive efforts on fewer individuals, giving them extra care to prepare them for survival to adulthood.

Humans fall in the latter category, generally producing no more than one offspring per year. At this rate, given the rigors of life encountered by early humans, there was little danger of overpopulation. Finding sufficient food while avoiding predators and accidents was a challenging task for most, and it is likely that very few of them died of old age.

Human numbers increased very slowly until only recently when they began to soar. A world population estimated at 300 million in 1 A.D. crept to 500 million in 1500, then doubled to 1 billion in 1804, to 2 billion in 1927, and to 4 billion in 1974. By the year 2000 the population exceeded 6 billion, with further increases yet to come, a surge that has paralleled the rise of technology. Principal contributors to this surge have been advances in medicine, less hazardous lifestyles, and improvements in food preparation and distribution, all direct results of our increasing technology in these areas. One can argue that extraterrestrial civilizations will experience a similar link between rising technology and rising population.

Such a sudden rise in world population can pose real problems for a civilization in terms of maintaining an adequate food supply for everyone, especially if the population growth remains unchecked. A civilization that is hard pressed to keep everyone fed will be less likely to spend its resources exploring space and searching for aliens, and may even suffer civil strife sufficient to subvert its technical status.

A larger population with an increasing concentration in crowded cities also makes the spread of communicable diseases much easier. This problem was evident in the cities of Europe during the 14th century when the bubonic plague killed about one third of the population within five years. Due to a lack of medical technology, the reasons behind the plague were not understood at the time, and treatments were generally ineffective. Since then, the causes of many diseases have been discovered, and considerable progress has been made in their prevention and treatment. The complete eradication of some diseases (such as smallpox) and the control of others through vaccinations led many optimistic observers to conclude that modern technology would eventually be able to provide humans with a world free of illness. So far, this does not appear to be the case.

The appearance of AIDS, Ebola, and other new diseases in the last half of the 20th century gives the distinct impression that humans are not really in control of this planet. Many old diseases are making a comeback or holding their own, due in large part to the high cost of drugs to combat them and the lack of health care for those afflicted. Newer, even more deadly diseases such as Ebola strike quickly and then retreat to their unknown abodes in the forests and jungles of Africa, giving researchers little chance to find cures. Meanwhile, our very efficient modes of transportation provide an excellent mechanism for diseases to spread rapidly across the globe, making no place on Earth truly safe from this peril. And to make matters worse, some nations of the world have supported research in biological warfare, deliberately trying to create biological agents that will menace the bodies of their enemies – and their friends, if they are not sufficiently careful.

Another problem with some of the microbes that cause diseases is their ability to evolve rapidly and/ or exchange genes to produce a more drug-resistant strain, which is then immune to the normal treatment. Overuse of antibiotics in treatment of diseases (or suspected diseases) unnecessarily exposes the omnipresent microbes to the antibiotic, giving those cells that already have some degree of resistance an excellent opportunity to become even more resistant. The less-resistant strains may indeed be eliminated by antibiotic treatment, but the more-resistant strains can survive to reproduce, passing their enhanced defenses on to future generations. The microbes – which have been in existence for over 3 billion years – are so good at this that we may eventually wind up with a host of untreatable diseases that keep the human population in check and limit its level of technology, if not its presence on the planet. Will extraterrestrials fare any better against their own lineup of nasty bugs?

In a variation on this theme, we might anticipate that human or robotic expeditions to Mars could bring Martian microbes back to Earth (if they exist). Such space bugs may be based on the same biology that we use and thus have the capability to thrive in our environment, perhaps unchecked by our natural defenses. Or the Martian bugs might be just similar enough to interact with terrestrial organisms, but different enough to be immune to our normal treatments. Naturally, precautions will be taken to prevent the human race from being wiped out by the Martian flu after rocks and soil from Mars are delivered to our planet for study. (And one hopes that any aliens who might be visiting Earth are taking similar precautions to avoid getting sick so far from home.)

Perhaps our increasing technology will be able to keep pace with the microbes after all, and ways will be found to prevent even the citizens of the poorest nations from succumbing to their attack. Perhaps every citizen of the planet will enjoy a long life and reproductive opportunities limited only by the need to hold the population at some reasonable level. What then?

The way the game of evolution is designed, anyone can play. Those players with attributes and skills that provide a competitive advantage will be more likely to survive and reproduce; those who are deficient in some way will be at a disadvantage and thus less likely to pass their genes on to future generations. For most of the existence of humans, we have evolved according to the rules of the game. Individuals that were smarter, faster, less susceptible to diseases, etc., were more apt to survive and reproduce. However, as civilizations advanced, it became possible (and often desirable) for those less-well-endowed individuals to survive with the assistance of others. The rise of technology allowed even more deficiencies to slip through the evolutionary cracks and be passed on to succeeding generations.

Today the human gene pool contains a host of characteristics that evolution probably would have discarded long ago (and perhaps did) because they produce individuals with inferior chances of surviving in the wild. As a common example, consider the millions of people who require corrective lenses for everyday tasks such as reading and driving. Although reading and driving skills were probably undervalued by primitive humans, the ability to clearly see an approaching predator or potential prey would have been highly prized. Presumably, those born with poor eyesight did not fare so well in the evolutionary derby. (Had he lived back then, the author would probably have been run over by a mammoth fairly early in life.)

Today we continue to evolve humans with bad eyesight, but rather than leaving them to their fate, we give them eyeglasses that allow them to lead nearly normal lives and produce offspring with similar problems, all thanks to the magic of technology. It should be a safe bet that the collective vision of the human population is getting worse, rather than better. The same can be said for many other deficiencies, all beneficiaries of our increased technological skills. The result is an overall deterioration of the human gene pool, insuring that successive generations will have a greater proportion of individuals who are less suited to the rigors of life in the wild.

Of course, our civilization's level of technology makes it unnecessary for *all* of us to be skilled hunters, fishermen, or farmers in order to acquire our next meal, but in the long run, genetic deterioration may prove to be detrimental to the human race. This is not to say that everyone with bad eyesight should be prevented from reproducing; such 'solutions' were attempted in the 20th century, to the displeasure of most of the world. But it may be that this is one of the traps technology lays for the unwary civilization, permitting very few of them to advance indefinitely up the evolutionary – and perhaps the technological – ladder.

A technological trap that has already raised its ugly head on this planet is the specter of nuclear war. War is nothing new to the human race, as any reading of history will demonstrate, but it has been only recently that weapons have been developed that threaten the existence of our entire species. Conventional weapons, whether stones, swords, arrows, bullets, or bombs, are capable of killing humans, individually or in large numbers, but once the blow has landed or the bomb exploded, the damage is done. Although one civilization may be able to eradicate a competitor with these tools, the victor is not apt to be in danger once the battle is won.

Nuclear weapons, on the other hand, have long-term effects that extend far beyond the destruction caused by the initial blast. Harmful emissions from the radioactive products of the explosion may continue for years after the event. Injection of such particles into the upper atmosphere permits the global spread of radioactive fallout, producing radiation sickness in individuals exposed to this invisible danger simply by being downwind from the target. Multiple detonations of nuclear devices would likely send enough ash into the atmosphere to block sunlight and generate a nuclear winter lasting for several years – long enough to inhibit crop production around the world and make life generally miserable for the 'lucky' survivors of the nuclear war.

It is hoped that such a scenario is not inevitable; the ending of the cold war has greatly reduced the likelihood of nuclear confrontation between the United States and the Soviet Union (which no longer exists), but there are still sufficient weapons of destruction to plunge the world into a nuclear holocaust if we are not careful. Furthermore, the spread of technology has made construction of nuclear weapons feasible for an increasing number of nations and terrorist groups, who might not be able to exercise appropriate restraint in their use of such weapons. As time goes on, the nuclear threat may again become critical.

Will alien civilizations also suffer from this particular danger? Will they have wars among themselves, or are humans unique in this respect? Without direct knowledge of *any* alien civilizations, it is difficult to say; however, one may argue that the evolutionary process that produced our species on Earth will be likely to work the same way on other planets. The individuals that survive to reproduce in any species are generally those that are better suited to living in their particular environments and playing out their designated roles as they compete for food and reproduction opportunities.

This competition takes different forms in different species. In humans, the winner may be the biggest, strongest, fastest, cleverest, or best at using technology – whatever provides a competitive advantage. The competitive spirit is obvious in humans; it is at the root of most wars as different factions compete for land or resources, and it forms the basis of most of our games and sporting events. Competition appears to be intertwined with the evolution of the human species, and it may well be equally ingrained in the cultures of intelligent extraterrestrials. If so, they too will have to deal with the peril of nuclear weapons and their own imminent destruction.

CIVILIZATION LIFETIMES

Clearly, there are *many* different reasons why a technical civilization should not be expected to last forever once it has developed. Assuming that extraterrestrial civilizations have indeed been formed in our Galaxy, the number of them that *currently* exist must depend on their various lifetimes. If lifetimes are generally short, technical civilizations will flare briefly into existence, only to wink out before encountering any others or advancing beyond the Kardashev Type I level. If civilizations can avoid the many pitfalls that await them and last as long as their suns remain stable, there could be aliens scattered all over the Galaxy just waiting to be discovered.

Selecting a value for the lifetime of an average technical civilization (L) to use in the Drake equation is not an easy task. We know of only one technical civilization (our own), but because it has not yet ended, we have no idea what its lifetime will ultimately be. We can only guess which of the many fates will be most likely to limit our reign over this planet, assess the degree to which these fates will be lethal, and estimate a time frame in which they might occur. As for alien civilizations, we cannot know exactly what dangers they might face, but if they have evolved around a star similar to ours, on a planet similar to ours, by processes similar to ours, using atoms and molecules similar to ours, they should be subject to many of the same limitations as we have listed for ourselves.

We can make a few estimates without too much difficulty; the Sun will last for a few billion years more, but our oil reserves will most likely have disappeared long before that. However, if alternative sources of energy and hydrocarbons can be developed, the exhaustion of our oil may not be so crucial. A large asteroid could land on Chicago in 50 years, or on the glacier that may be covering Chicago in 50,000 years. The Yellowstone supervolcano may erupt some time in the next millennium, by which time humans may be living underground to shield themselves from ultraviolet radiation streaming through a global ozone hole. Perhaps we will be clever and fortunate and survive for another million years; or perhaps the microbes will find a way to poison us all by then, leaving the planet in the care of a completely different dominant species.

Obviously, the subject of technical civilization lifetimes is fairly wide open, leaving the reader free to choose whatever seems to make sense. Because there is so much uncertainty with this factor, one should take care to try to select an appropriate lifetime for the chosen hazard; for example, asserting that the Sun will burn out in the next thousand years is not very realistic, based on our present understanding of stellar evolution. The value for this factor is independent of all the other values in the Drake equation: there may be many civilizations with very short lifetimes or only a few, with relatively long lifetimes — any combination is possible.

Selection of this final factor will allow the Drake equation calculation to be completed, giving the number of technical civilizations in our Galaxy today. Although the same evidence and arguments for the existence of extraterrestrial civilizations are available for everyone to consider, each person who uses the Drake equation may arrive at a different number. The next chapter will attempt to reconcile this apparent contradiction.

MAIN IDEAS

- Some intelligent species that evolve will be unable to develop technical civilizations, due to deficiencies in their body types; others may be restricted by lack of appropriate planetary surfaces and/or access to necessary materials, such as metals or fossil fuels.

- Although different levels of technology (such as Kardashev civilizations) can be defined, it is not obvious whether it is actually possible for a civilization to upgrade from one level to the next.

- The lifetimes of technical civilizations may vary widely as there are numerous hazards to the existence of any civilization, and these hazards occur on different time scales that range from a few decades to billions of years.

- If the lifetimes of technical civilizations tend to be relatively short, then few of them may overlap in time even though many may arise in the Galaxy.

KEYWORDS

asteroid impacts	Dyson spheres	Kardashev civilizations
mass extinctions	meteor	meteorite
meteoroid	ozone layer	subduction zone
supervolcanoes	tsunami	Tunguska event

LAUNCHPADS

1. Could a planet possibly produce a technical civilization without the benefit of a supply of fossil fuels? How might such aliens overcome this problem?

2. By what steps could a technical civilization possibly make the huge leap from Kardashev Type I to Type II?

3. Suppose that human lifetimes could be increased to 10,000 years. How would this change affect the ability of our technical civilization to survive on Earth?

REFERENCES

Breuer, Reinhard A. 1982. *Contact with the stars.* San Francisco: W.H. Freeman.

Chapter 11
AN ABSENCE OF EVIDENCE

*I*n which the status of the evidence for extraterrestrial civilizations is reviewed, other similar phenomena are presented for comparison, numerous hypotheses consistent with the evidence for alien civilizations are provided, and guidelines for selecting a logical hypothesis are suggested.

By this time, the reader should be ready for some good, solid answers. Do aliens exist in the Galaxy or not? Are they relatively nearby or too far for us to ever hope to contact? Have they visited Earth? Unfortunately, the answers to most of these questions are the same: we really do not know for sure. However, with the knowledge acquired by reading this text – and perhaps several of the other fine books on the subject – the reader should at least be able to form some good, solid hypotheses.

Hypotheses are explanations of observations, and any hypothesis that is proposed for this particular discussion should attempt to be consistent with the observational evidence concerning alien civilizations in the Milky Way. What is the current status of such evidence?

EXTRATERRESTRIAL TESTAMENTS

Although much has been written and claimed about the idea of 'ancient astronauts', there is no *convincing* evidence that Earth was ever visited in ancient times by aliens who inspired the construction of pyramids and other monuments or instructed primitive humans on how to create such things. So far, these wonders are quite reasonably explained as the products of the brains and brawn of local human cultures. We may not understand every detail of pyramid construction, but it does not appear to be anywhere near as difficult as interstellar space flight. If the aliens really wanted to help us by lending their expertise to our more challenging technical projects, one might think they would have showed up to help get us to the Moon.

As for current visits by extraterrestrials, despite numerous sightings of UFOs by the general public and scattered revelations by humans claiming to be survivors of alien abductions, the scientific community does not appear to be convinced. Anyone can see a strange light or object in the sky – there are plenty of them up there. However, most scientists are reluctant to rely on extraterrestrial spaceships as the default explanation for such phenomena, due to the speculative nature of their existence. Similarly, tales of abductions by aliens offer little that distinguishes them from dreams or other fabrications of the mind. As much as those claiming the authenticity of these abductions may believe in them, there is still insufficient tangible evidence to persuade the majority of scientists of the reality of such events.

Searches for electromagnetic waves from intelligent aliens have fared no better. Although radio astronomers have occasionally found interesting signals, none have been confirmed as actual messages from extraterrestrials. Of course, the coverage of stars and frequencies has not been anywhere near complete; there are *many* places to look, and the searches to date have attempted to focus on the most likely possibilities. As the searches continue and improvements are made to detection equipment, the chances of receiving any alien signals that may be present will increase. But so far our searches have produced no clear evidence of extraterrestrials.

One might say that when it comes to evidence, extraterrestrials are in the same league as several other strange phenomena that have not had their existences adequately established. The most obvious of these include the Loch Ness monster (Nessie) and the hairy humanoid creatures known variously as Bigfoot, Sasquatch, Yeti, etc. There are many claims of sightings of each of these – grainy photographs of Nessie, casts of Bigfoot's footprints, and so on – but never is there sufficient evidence of these creatures to convince the scientific community of their reality. Nor are there many scientists willing to spend their time researching such marginal topics. It seems that for every tale of a close encounter there is a sugges- tion of a hoax, and few researchers are eager to champion such causes when there are legitimate scientific discoveries to be made elsewhere.

Other questionable characters could be mentioned here. Literature provides us with quite a collec- tion of gnomes, leprechauns, elves, fairies, goblins, trolls, dragons, wizards, witches, ghosts, zombies, werewolves, vampires, demons, devils, angels, gods, and the like. Most of these do not get as much seri- ous attention as they probably did in pre-technology days, when the world was a much more mysterious place. Could some of these have originated from encounters with extraterrestrials (providing possible evidence of alien visits!), or are they simply products of the human brain, created to entertain, frighten, or warn a less-educated public than we find in our current modern society?

While it would be difficult to produce much hard evidence of trolls today, there are a few mythical char- acters that *do* appear to leave us some clues to their reality. Santa Claus, the Easter Bunny, and the Tooth Fairy have all made careers out of leaving tangible evidence of their visits, in the form of presents, candy, and cash. Countless children have performed scientific experiments to validate their trust in the existence of these bene- factors, and in many cases their results have matched – or sometimes exceeded – their initial predictions.

In the particular case of Santa Claus, there is considerable evidence that could be used to show that he is, in fact, real. Photographs of Santa abound, and his appearance is familiar to all. There is general agreement on his humanoid form and the costume he wears. Near the time of his regular visits, he can often be seen in shopping malls, where some report actually having sat on his lap, conversing with him (he seems to have no difficulty with our language). His principal mode of transportation is well known, although the details of its propulsion system are not completely understood. He appears to have a variety of magical powers that allow him to move freely through barriers and travel rapidly from place to place. He is said to be omnipresent and omniscient, qualities not normally found in benevolent terrestrial life forms. In short, a strong argument could be made for Santa's being both real and non-human. Perhaps Santa Claus is an extraterrestrial!

Of course, there will always be some who deny the reality of certain phenomena, ignoring what is taken by others to be overwhelming evidence for these things. And there will often be those who cling to the truth of an idea, despite overwhelming evidence to the contrary. These are extreme positions, normally not particularly useful within the framework of science, where hypotheses are to be judged on evidence. In cases where solid evidence is completely lacking or cannot be agreed upon – such as the current question of extraterrestrial civilizations – the urge to cling to one's beliefs may be even stronger. Many people will rely on 'gut feelings' as they form their opinions about alien existence. Whatever method is used, the field is still wide open, due to our inadequate supply of evidence.

A HOST OF HYPOTHESES

Lack of evidence of alien civilizations does not prevent us from formulating hypotheses to explain the situation. In fact, quite a variety of different explanations can be developed, each of them perfectly consistent with our current observations. Some of these hypotheses are very elaborate while a few are quite simple. Although we do not know which, if any, of these ideas is correct, we *can* say that most of them are undoubtedly wrong.

Our presence in the Galaxy as a technical civilization gives us a reference point from which to work. The hypotheses we develop can be grouped into three general categories, which are based on the number and/or technical level of the civilizations envisioned by each hypothesis:

- **Category #1:** We are alone – the only technical civilization in the Galaxy.

- **Category #2:** There are other technical civilizations like us in the Galaxy, but none that are significantly more advanced.

- **Category #3:** Advanced technical civilizations that are far beyond our level of technology exist in the Galaxy.

Let us examine each of these categories in turn, to explore the wide range of ideas available to explain the current absence of compelling evidence for extraterrestrial civilizations.

Category #1: We are alone – the only technical civilization in the Galaxy.

This category presents the simplest explanations of all: we have no evidence of other civilizations because they do not exist. For most of these hypotheses, a single factor is sufficient to prevent civilizations from developing. Our presence here is then regarded as either an anomaly or a very transient state of the Galaxy.

- **Hypothesis #1.A:** We are alone because no other suitable planets exist in the Galaxy.

We have yet to observe *any* terrestrial planets outside our solar system, let alone any that resemble Earth. The requirements on the evolution of the temperature, atmosphere, and oceans on a planet that would develop life leave essentially no margin for error. The combination of conditions and sequence of events that made life on Earth possible has not occurred anywhere else in the Galaxy, making Earth a unique world among all the planets of the Milky Way, however numerous they may be.

- **Hypothesis #1.B:** We are alone because starting life is actually very difficult.

Even given the proper planetary conditions, life is a long shot, requiring a highly improbable (and still unknown) sequence of events to occur at the molecular level. The formation of life on Earth was a fluke, an action not likely to be repeated, even on other Earth-like planets. As a result, life of *any* kind is rare in the Galaxy.

- **Hypothesis #1.C:** We are alone because fabricating complex life forms from single cells is actually very difficult.

Simple life develops easily, given the chance, and while there are numerous planets populated by single-celled life, higher life forms have managed to evolve only on the Earth. Throughout the first 3 billion years or so of terrestrial life, complexity was essentially nonexistent; only in the last few hundred million years has evolution managed to produce the complex plant and animal life we experience today. Very unusual conditions caused or encouraged individual cells to work together toward the formation of complex organisms on Earth, and similar conditions are extremely rare – perhaps nonexistent – in the Galaxy.

- **Hypothesis #1.D:** We are alone because the evolution of intelligent life is actually very difficult.

Although we may sometimes regard ourselves as the pinnacle of evolution, we are really *not* the ultimate goal of this process – for evolution has no particular goals. Evolution produces a variety of species, and those equipped with the best survival mechanisms will be successful at reproducing and continuing the species. Intelligence is one of many survival mechanisms, but whether it is the *best* depends on the criteria being used to judge the contest. In terms of numbers of species, numbers of individuals, or length of time on the planet, intelligent species rank far behind many others. In terms of the ability to dominate – and perhaps destroy – other life forms on the planet, intelligent species (especially humans) do seem to have an edge. Of course, if the dominance of humans eventually results in their own destruction (see Chapter 10), then evolution will have worked its magic once again.

It is not obvious that evolution will *necessarily* result in intelligent species; there are plenty of other species that fill the various ecological niches without too much reliance on this characteristic. If evolution really did move species in the direction of increasing intelligence, we should probably see a greater diversity of intelligent species on Earth – intelligent insects, intelligent birds, intelligent mollusks, etc. But it may be that intelligence at the level of ours can only develop in concurrence with certain body types, perhaps those that permit the use of a variety of tools and the generation of enough different sounds to support a significant vocabulary. A large brain may not be sufficient to produce what we consider to be an intelligent species.

Given enough time, evolution can create a wide variety of species on a trial basis, but an intelligent species still might not be able to compete successfully with other species for dominance of the planet. For example, humans had the good fortune to evolve on Earth at a time when they had minimal competition from numerous large, fierce predators, such as dinosaurs. The success of intelligent species may well be determined by good timing and good luck, which are not necessarily inevitable on every life-bearing planet.

- **Hypothesis #1.E:** We are alone because intelligent species rarely produce technical civilizations.

Intelligent life forms with appropriate manipulative appendages and suitable environments are extremely rare. Those intelligent species that cannot construct or use radios, radio telescopes, or other electronic devices will be unable to achieve our technical status. Alternatively, planets with sufficient access to the materials of technology are rare. If fossil fuels have not formed, if metal ores are not available on the surface of a planet, then intelligent species will be unable to utilize them to develop a technical civilization. The Galaxy could be well populated with intelligent species that have no hope of producing a technical civilization.

- **Hypothesis #1.F:** We are alone because technical civilizations that do form cannot maintain their existence for very long.

Technical civilizations may have the misfortune to be wiped out by asteroid impacts, volcanic activity, ice ages, or other relatively frequently occurring natural disasters that are beyond their control. Or they might use up their supplies of fossil fuels and other raw materials required to sustain their technical activities. Technical civilizations could also tend to destroy themselves, through nuclear wars, overpopulation, pollution, diseases, etc. There are so many different fates that can ruin a technical civilization that long-term survival has a very low probability.

Category #2: There are other technical civilizations like us in the Galaxy, but none that are significantly more advanced.

This category allows the development of numerous Kardashev Type I civilizations in the Galaxy, but presumes that advancement beyond this stage has not occurred. As Type I civilizations are not capable of interstellar space travel, these hypotheses preclude visits from any aliens. Any knowledge that we might have of them would probably have to come through radio contact.

- **Hypothesis #2.A:** There are other Type I technical civilizations in the Galaxy, but they are too far away for us to have established contact with them during the brief lifetime of our own technical civilization.

They may not be aware of our presence because our radio signals have not yet had time to reach them. Or they may have received our signals, but decided not to reply. If they have replied, their signals have not reached us as yet simply because their star is too far from our Sun.

Another possibility is that these other civilizations do not have active radio search programs or at least not very effective ones. Our signals could have reached their planets without detection by their civilizations, just as we may have missed signals beamed in our direction by extraterrestrials.

- **Hypothesis #2.B:** Another Type I technical civilization exists relatively nearby, but they are not sending any radio signals to reveal their presence.

Although civilizations such as ours are quite capable of beaming radio signals to nearby stars, they may not necessarily be doing so. Such programs require an investment in time and resources that an alien civilization might be unwilling to make. They might not even be listening for signals and thus would be unaware of our presence, despite being very close to us.

On the other hand, they might be listening for signals and could have received some of ours, but they may be too cautious to reply to such a close neighbor. It would be difficult for them to judge our level of technology and our intentions solely on the basis of our radio leakage, and they may decide to play it safe and wait for more information.

Category #3: Advanced technical civilizations (ATCs) that are far beyond our level of technology exist in the Galaxy.

This category presents the most complicated explanations of all. Each must explain why, despite the existence of advanced civilizations (at the level of Kardashev Type II or Type III), we have not yet managed to detect them. Because such civilizations should be engaged in some fairly major projects within the Galaxy, we would expect them to have some difficulty keeping their activities secret, or at least making them appear to be the result of natural forces. (The existence of ATCs does not preclude the presence of other Type I civilizations in the Galaxy; the hypotheses in the second category can still be applied to these civilizations.)

- **Hypothesis #3.A:** ATCs are not aware of us; if we should discover them, it would have to be by accident (which has not happened yet).

ATCs might not be searching for technical civilizations such as ours. If this is the case, they would not be sending out space probes to locate us or radio signals for us to intercept. Or perhaps they do not even use radio waves for communication, but rather some more advanced method that is undetectable by us. As a result, our present search techniques are directed towards nonexistent transmissions.

Or it may be that ATCs *do* use radio signals for communication, but we have not detected any as yet. They may carefully control the direction their signals are beamed, and none have been sent in our direction. Or they may use a radio frequency or bandwidth that we have not tried or cannot use, perhaps because it is heavily employed in commercial applications here on Earth (CB radios, garage door openers, etc.).

The ATCs could have unknowingly transmitted signals in our direction, but we were not listening for them at the time they arrived. Or the ATCs could be so far away that the signals that reached Earth have been too weak for us to detect. We may have picked up occasional radio signals from ATCs but were unable to recognize or confirm them.

ATCs could be sending out space probes for various purposes, but none of them have come here. They may not intend to conduct a thorough search of the Galaxy, perhaps only searching until they encounter another civilization. Or they could be searching through the entire Galaxy but have not had the time to examine our solar system yet.

It is also possible that space probes from an ATC came to Earth long ago and then moved on, before humans had even developed. They could have already crossed Earth off the list of inhabited worlds.

- **Hypothesis #3.B:** ATCs are aware of our presence.

The ATCs could be trying to contact us but do not know quite where we are. They may have intercepted our radio signals but lacked enough information to pinpoint our location. As a result, they do not know which direction to beam a message or send a probe, and their radio signals, broadcast in many directions, have been too weak for us to detect.

Alternatively, the ATCs might have a pretty good idea where we are, based on many years of receiving our radio leakage. They could be sending radio messages in our direction, but there might not have been time for the signals to travel here yet. Or the signals could have arrived, but we did not receive them. It is also possible that we actually received their signals but did not recognize them as messages.

If the ATCs know where we are, they might also decide to launch space probes towards us. If this is the case, either the probes have not arrived yet, or they are here but we have not seen them. Or possibly, we *have* seen them, but they were just one of many UFO reports.

If the ATCs have mastered interstellar space travel and are not too far away, they may decide to come in person, in which case either they are still en route, or they have already arrived. If they are here, we might not have noticed them yet, although this seems odd for a civilization that is deliberately trying to contact us. Or we may have noticed their spaceships, but again, they were written off as UFOs. It is even possible that the aliens are here among us, but are so well disguised that we do not notice them. Why they would successfully travel all the way to Earth only to fail in their efforts to contact us must be somewhat baffling for proponents of this hypothesis. Given sufficient time, an ATC should be able to accomplish such a task if they are so inclined.

Perhaps ATCs do *not* desire contact with civilizations such as ours (or with ours in particular), even though they know of our existence. They may have no interest in us; after all, as a Type I civilization, we would have little to offer them in terms of technology. Besides, if ATCs exist, Type I civilizations may be commonplace and not considered worthy of close examination.

On the other hand, the ATCs may indeed have an interest in us, but do not want us to know about them. They may be observing us in secret, perhaps wanting to see just what type of life we are before admitting us into the association of technical civilizations that coordinates operations within the Galaxy. There may be minimum requirements for us to meet and possibly a trial membership period, but secret handshakes are unlikely.

Perhaps we do not yet measure up to their standards. They could be here on the Earth trying to help us along, preventing us from destroying ourselves with our newfound technology. They may be masquerading as humans while they participate in – or interfere with – our everyday lives. However, given the apparent improbability of extraterrestrials with humanoid forms, especially those capable of adapting easily to Earth's gravity and atmosphere, it is difficult to imagine how aliens could blend in with our population without attracting undue attention.

A more radical explanation proposes that aliens are responsible for our very presence here, having 'seeded' the Earth with life long ago. Since then, they have been secretly monitoring the planet to see what develops, perhaps without interfering with the results. They could have planted similar 'seeds' on other planets throughout the Galaxy, performing an extensive laboratory experiment running for hundreds of millions of years. Just *why* they would launch such a lengthy project is unclear, at least to those of us with individual lifetimes of only a century or so. A more plausible variation on this theme would have the Earth being seeded accidentally, by alien spacecraft that landed before life was well established here. This contamination could have destroyed any existing life forms and taken over the planet, ultimately resulting in our current biological system.

An even stranger idea supposes that the ATCs in the Galaxy regard our civilization as we regard the wild animals on the Earth – curious beasts to watch and study but clearly not up to our level of intelligence, ability, and power. In this 'Cosmic Zoo' hypothesis, Earth (or the solar system) serves as a natural 'cage' for the human civilization, keeping us safely quarantined, away from the more independent ATCs. We may be visited and/or studied by scientists or other citizens of the ATCs, but for their safety – and perhaps ours – we are not to be allowed to roam freely through the Galaxy.

Another possibility involves the role of religion in nearly every human culture around the world. The abundance of religions based on the existence of superior beings (often dwelling in the sky) that serve to guide the daily conduct of humans could be taken as evidence of the existence of extraterrestrials. Alien astronauts from ATCs would have undoubtedly appeared to early humans to have magical powers that allowed them to work miracles to impress their terrestrial hosts. Their mode of transportation and their communication methods would have added to their omniscient, omnipotent reputation and allowed the spread of their legend to all corners of the inhabited world. The gods of the various terrestrial religions could be extraterrestrials from ATCs, or they could have originated as a result of visits by these aliens. On the other hand, it may be more accurate to say that the human mind seems to have a propensity to construct religions, a tendency that may exist because it had some actual survival value in human evolution.

The above discussion of hypotheses is not meant to be all-inclusive; there are undoubtedly numerous other possibilities that might be considered, but those described here should cover most of the present favorites. The real question for the reader, of course, is how to choose *which* hypothesis is the most likely explanation for the current absence of evidence for extraterrestrials.

WHAT TO BELIEVE?

Forming an opinion about extraterrestrial life is not exactly a life-or-death matter. Most people can probably get through their entire lives without ever giving this topic a second thought, and in all likelihood, the correctness of anyone's opinion is not apt to be determined in the near future, if ever. The real value of this exercise is the practice gained in analyzing a collection of observations and opinions and trying to determine which of them make the most sense. Our everyday life is full of more mundane situations that compete for our attention and demand our common sense. Problems such as this one provide exercise for the brain and keep it in optimal working order – a good idea in case the aliens do appear on our doorstep some day.

Unfortunately, there are very few guidelines to follow. If there were many, opinions would probably tend to converge, and there would be far less uncertainty on this issue. The reader should already have *created* one guideline: the insertion of carefully chosen values into the various factors of the Drake equation will have produced a personal estimate of the number of technical civilizations in the Galaxy at present. The reader's subsequent choice of a hypothesis should certainly be in harmony with the results of his or her own Drake equation. For example, it would make little sense to calculate that 10,000 civilizations exist in the Milky Way and then select a hypothesis from Category #1 (which presumes that we are alone). Even with this constraint, there is still plenty of leeway in the choice of hypotheses, reflecting our lack of evidence that could help to limit the possibilities.

This situation is not unique; scientific investigators are often faced with a bewildering array of possible explanations for some phenomenon and very little evidence upon which to base a decision. In such cases, they may turn to a bit of advice from a 14th century monk, William of Occam (or Ockham), who made frequent use of a principle that is now widely known as **Occam's** (or Ockham's) **razor**:

Non sunt multiplicanda entia praeter necessitatem.

This can be translated into English as "Entities are not to be multiplied beyond necessity," which can be interpreted as "Any hypothesis should be shorn of all unnecessary assumptions," or in this case as "If two hypotheses fit the observations equally well, the one that makes fewer assumptions should be chosen."

Such reasoning is quite common in the scientific process. There can be numerous hypotheses available to explain an event or an object, some of them relatively simple and others more complex. Scientists usually look for a simple explanation first because such theories are easier to construct and generally are more broadly applicable. Additionally, the truth of any hypothesis depends on the concurrent truth of the assumptions upon which it depends; the more assumptions that are involved, the less likely that they will *all* be true.

As an example, consider a strange light seen moving silently and swiftly across the night sky. Such an object might be explained by some as an extraterrestrial spaceship, and by others as a meteor. Which of these requires fewer assumptions?

Meteors are common phenomena, occurring every day as millions of bits of rock collide with the Earth and burn up in the atmosphere. Sighting a meteor is not an unusual experience and in fact is to be expected if one watches the sky at night. On the other hand, sighting an extraterrestrial spaceship in the sky would require that (1) an alien civilization exists, (2) the aliens have mastered interstellar space travel, (3) the aliens have had sufficient motivation, resources, and time to come to Earth, and (4) the aliens fly through our skies at night with lights on their spacecraft, but are otherwise quite difficult to detect. Had Mr. Occam been faced with this choice of explanations, he clearly would have favored the meteor.

What does Occam's razor tell us about the existence of extraterrestrial civilizations? Temporarily ignoring any results from the Drake equation, we could compare our three categories, being on the lookout for any necessary assumptions. Each category explains the apparent absence of evidence of extraterrestrials in a different way. Category #1 hypotheses say that we have no evidence of any extraterrestrial technical civilizations because they do not exist. Category #2 hypotheses say that we have no evidence of aliens even though Type I civilizations do exist. Category #3 hypotheses say that we have no evidence of aliens even though advanced technical civilizations do exist.

Of these three groups, the simplest must surely be Category #1. If no civilizations capable of making contact with us exist, then there is little else to be explained, especially with the variety of reasons available to account for their absence. Further, the same reason need not be responsible for the lack of technical civilizations on every planet; several different hypotheses in this category may combine to explain the overall scarcity of such civilizations.

On the other hand, both of the other categories require additional assumptions about the distance, technical capabilities, motivation, and/or intentions of any alien civilization that is assumed to be present in the Milky Way. The greater the number of technical civilizations that exist, the less likely it is that any single hypothesis will explain why we are unaware of all of them. And if a great many of these civilizations – presumably with quite diverse capabilities and motivations – actually do populate our Galaxy, the likelihood that they will all manage to avoid detection by us would seem to be rather small.

Some may not be comfortable with this conclusion; they may find the prospect of our being the only technical civilization in the Milky Way to be less interesting, perhaps frightening, or simply unacceptable. They may argue that in the vast reaches of our Galaxy, among the hundreds of billions of stars, there must surely be other civilizations like ours. Perhaps Occam's razor is not applicable to this problem.

But if there really are other technical civilizations, then this should have been indicated by the results of the Drake equation. A person who calculates a large number of technical civilizations could still use Occam's razor to choose among the many hypotheses that are consistent with such a result. Occam's razor does not preclude any particular scenario; it is merely a tool that may help to point us in the direction of the truth – as long as the truth is not too strange.

With our meager base of observations concerning aliens, there is still plenty of room for diverging opinions. As far as the evidence is concerned, those who contend that we are the only technical civilization in the Galaxy are on equal footing with those who believe the Milky Way is crawling with them; either hypothesis could be consistent with our current lack of solid evidence. However, this need not mean that all hypotheses are equally likely – we may not be able to accurately predict the next roll of a pair of dice, but we can certainly say that rolling a seven is more probable than rolling a twelve. Unfortunately, determining the probabilities for the existence of extraterrestrial civilizations is far more difficult than analyzing any throw of the dice. Despite improved theoretical models of stars, planetary systems, and evolving planetary atmospheres, we are unlikely to figure this puzzle out without a lot more observational data.

WHAT NEXT?

As indicated earlier, the scale of this problem is so great that we will never have a complete solution, although we will undoubtedly continue to accumulate more pieces to the puzzle. Detection of extrasolar planets will probably improve to allow us to identify *terrestrial* planets orbiting nearby stars. We may be able to obtain spectra of these planets that would hint at their atmospheric compositions. An atmosphere rich in oxygen could be an indication of life on the planet; the faint spectral signature of an *artificial* substance such as a chlorofluorocarbon could even reveal the presence of a *technical* civilization, although such a pollutant could linger in the atmosphere long after the civilization had died out. And of course, continued searches for extraterrestrial radio signals may eventually turn up an unmistakable beacon or message announcing the existence of another civilization. We may find out – within our lifetimes – that we are not alone.

What if we do locate another civilization? What effect would such a result have on the search for extraterrestrials? The discovery of just one alien civilization should allow us to rule out *all* of the hypotheses of Category #1; it should also provide some hope that our technical civilization is not necessarily destined to fail in the near future. Funding for larger telescopes, expanded radio searches, and related research projects would probably be easier to obtain, as the prospect for positive results should appear considerably brighter. And of course, substantial revisions would be in order for astronomy books and courses that delve into the topic of "Life in the Universe" (not to mention the chaos generated as exobiologists finally are given some observational data to study).

Effects on the general population are a bit more difficult to gauge. Some think that finding extraterrestrial civilizations will change *everything* here on Earth, implying that somehow such a discovery will bring humans together, ending our petty international squabbles and making us realize that we must work together for the survival of our species rather than fighting among ourselves. This is a nice idea, but it is unlikely to happen.

The aliens we discover are not apt to pose any *immediate* threat. Civilizations are more likely to learn of each other through radio signals than through personal contact. Without an alien on our doorstep, humans will have no particular incentive for global cooperation.

If the aliens are capable of interstellar space travel and do decide to pay us a visit, their trip will probably last a long time – several decades or centuries at least, given the distances of the nearby stars. But humans tend to be relatively shortsighted, thinking ahead a few decades at most and seldom much beyond the end of their own lives. Preparations for the arrival of an alien expedition will probably be made at the last minute, long after we are gone; in the meantime, life will probably go on as usual once the initial headlines have died down.

And what if we do *not* find evidence of another civilization? What if we continue to fail to detect interesting terrestrial planets or alien radio signals? At what point will we finally conclude that we are in fact alone, that our presence in the Galaxy is a fluke, and that our civilization has no hope of contacting any intelligent extraterrestrials who might be able to provide us with solutions to our many problems? When does our "absence of evidence" actually become "evidence of absence"?

These are questions without any real answers, for while some people have already drawn the conclusion that extraterrestrial civilizations do not exist, others refuse to give up hope. As long as there are individuals and organizations willing to fund the SETI radio searches, and scientists can still be found to conduct them, the effort to find other civilizations will probably continue.

There are many who consider the thought that we might be alone in the Galaxy to be an absurdity; there are others who are quite convinced that we are, in fact, unique. The arguments and facts presented in this book could be followed to either conclusion, depending on the reader's point of view. Regardless of the actuality, it is probably safe to say that the existence of life has had no significant effect on the overall structure and operation of either our Galaxy or the universe. Either of these systems can function equally well with or without life.

Our own presence in the universe, while quite fortunate for us, is really unnecessary. The universe owes us nothing but tolerates us anyway – at least it has so far. We exist today in a tiny corner of the universe that is tailor-made for our species – or rather, our species was tailor-made to fit this particular corner.

Our favorable situation will not last forever. The universe, the Galaxy, the Sun, the Earth, and humans are all evolving on different time scales. Eventually our species will find itself out of sync with its own little corner, and we will be unceremoniously scratched from the roster of technical civilizations – brought down by some wayward asteroid, ice sheet, microbe, or pollutant. Will our disappearance be noticed by alien civilizations? Will we be mourned? Will we eventually be replaced by a new-and-improved intelligent terrestrial species?

Unanswered questions such as these may seem largely irrelevant to our current daily lives – as some consider the entire idea of extraterrestrial life to be. But this subject provides much more than just interesting conversational topics. The real value of our quest for cosmic companionship is in the mind-stretching exercises it requires. Understanding the arguments on both sides involves the digestion of a significant amount of basic scientific knowledge, and formulation of one's own ideas requires logical thinking and some degree of creativity as well. The mental processes used in developing these ideas are those that have enabled our civilization to become technical. Whether the opinions we generate are right or wrong, the mental exercises performed in arriving at them will ultimately help to insure that we maintain our own technical status. Intellectual pursuits such as the search for intelligent extraterrestrials are absolutely necessary for the survival of our own civilization.

MAIN IDEAS

- After years of UFO reports and a host of hypotheses attributing a variety of strange human experiences and unexplained phenomena to alien visitors, there is still insufficient evidence to convince the scientific community that extraterrestrials have arrived at Earth.

- Without any direct evidence of extraterrestrial civilizations, we are left with a number of different scenarios that could explain our situation; these range from our being alone in the Galaxy to our being one of perhaps many similar civilizations to our being one of the lesser civilizations in a Galaxy populated by one or more advanced technical civilizations.

- Occam's razor, which advises us to choose the hypothesis that involves the fewest assumptions, may provide some guidance in selecting the explanation that is closest to the truth.

KEYWORDS

alien abductions   **ancient astronauts**   **Occam's razor**
unidentified flying object (UFO)

LAUNCHPADS

1. Suppose that we had solid evidence that extraterrestrials had in fact visited Earth in the time of the ancient Egyptians. What hypotheses for extraterrestrial civilizations could be proposed that would be consistent with this evidence?

2. Are intelligent extraterrestrial beings apt to be much larger or much smaller than humans? Could sightings of gnomes, fairies, giants, and such have anything to do with aliens?

3. Some people have claimed that NASA photos of the Martian surface reveal a huge face, sculpted by aliens. What assumptions must be made in order for this hypothesis to sound plausible?

Appendix

LARGE NUMBERS AND SMALL FRACTIONS

In the search for extraterrestrial life in the universe, we will frequently encounter both large numbers and small fractions. It is best to be familiar with both types of numbers.

Here are some examples of basic nomenclature:

$$
\begin{aligned}
1,000,000,000 \quad &= 10^9 = \text{one billion} \\
1,000,000 \quad &= 10^6 = \text{one million} \\
1000 \quad &= 10^3 = \text{one thousand} \\
100 \quad &= 10^2 = \text{one hundred} \\
10 \quad &= 10^1 = \text{ten} \\
1 \quad &= 10^0 = \text{one} \ = 100\% \\
0.1 \quad &= 10^{-1} = \text{one tenth} = \text{one in ten} = 10\% \\
0.01 \quad &= 10^{-2} = \text{one hundredth} = \text{one in a hundred} = 1\% \\
0.001 \quad &= 10^{-3} = \text{one thousandth} = \text{one in a thousand} = 0.1\% \\
0.000001 \quad &= 10^{-6} = \text{one 1 millionth} = \text{one in a million} = 0.0001\% \\
0.000000001 \quad &= 10^{-9} = \text{one billionth} = \text{one in a billion} = 0.0000001\%
\end{aligned}
$$

[Beware: the British consider a billion to be a *million* million, rather than a *thousand* million as Americans do.]

Large numbers are needed to discuss vast quantities of objects, such as the number of sand grains on a beach. Small fractions are encountered when considering highly improbable occurrences, such as the fraction of sand grains that are *blue*.

Suppose we are interested in the *number* of blue sand grains in a bucket of sand. We could determine this number by sifting through all of the sand grains in the bucket and counting the blue ones, but that would take a long time. Or we could obtain the number of blue grains by estimating the total number of sand grains in the bucket and multiplying it by the *fraction* of blue sand grains. This would be an estimate – and it would only be as good as our estimates of the two numbers we used – but it would certainly be quicker than counting.

For example, consider two possible sets of estimates:

	Total sand grains	x	Fraction of blue grains	=	Number of blue grains
A	1 billion	x	1 in a million	=	1000
B	100 million	x	1 in 10 billion	=	0.01

If our estimates of case A are correct, we should be able to find about 1000 blue grains of sand in the bucket. If our estimates of case B are correct, we should expect to find only one blue grain in every *100 buckets* of sand.

Similarly, this book is concerned with the abundance of a particular type of entity: extraterrestrial civilizations. And as in the case of the blue sand, the abundance of extraterrestrial civilizations will be difficult (if not impossible) to determine by looking at each potential site in the Galaxy. We will be better off to try to make reasonable determinations of both (1) the *number* of potential sites for extraterrestrial civilizations in the Galaxy, and (2) the *fraction* of sites that actually support extraterrestrial civilizations. The product of these two numbers – a large number and a small fraction – will give us our estimate.

Determination of the large number is usually not too difficult; it is the estimation of the small fraction that will give us the most problems. In some cases our small fraction will be a *probability* [see Chapter 2] that can be calculated.

PROBABILITY AND CARDS

A normal deck of cards contains 52 cards, consisting of 4 suits:
Spades and clubs (which are black), and hearts and diamonds (which are red).
Each suit contains 13 cards: A, 2, 3, 4, 5, 6, 7, 8, 9, 10, J, Q, K.
J, Q, and K are face cards.

How many red cards are in the deck?	26
How many diamonds are in the deck?	13
How many face cards are in the deck?	12
How many queens are in the deck?	4
How many queens of diamonds are in the deck?	1

What is the probability of drawing a card at random from the full deck and having it be

a red card?	$26/52 = 1/2$
a diamond?	$13/52 = 1/4$
a face card?	$12/52 = 3/13$
a queen?	$4/52 = 1/13$
the queen of diamonds?	$1/52$

Suppose we mix 1000 decks together to make a 'super deck'.

How many cards are in this super deck?	52000
How many red cards are in this super deck?	26000
How many diamonds are in this super deck?	13000
How many face cards are in this super deck?	12000
How many queens are in this super deck?	4000
How many queens of diamonds are in this super deck?	1000

What is the probability of drawing a card at random from the super deck and having it be

a red card?	$26000/52000 = 1/2$
a diamond?	$13000/52000 = 1/4$
a face card?	$12000/52000 = 3/13$
a queen?	$4000/52000 = 1/13$
the queen of diamonds?	$1000/52000 = 1/52$

The same probabilities apply to both large and small decks because the large deck is constructed the same way as the small one. Because of this similarity, we can use our knowledge of normal decks to make estimates of the super deck. However, if the additional 999 decks that were used to make the super deck had been different from our original deck, then our estimates of properties of the super deck could be quite in error.

Our solar system is perhaps one of many in the Galaxy. If it is typical, we could use our knowledge of the Earth and nearby planets to estimate properties of other planetary systems in the Galaxy. But if our system is unique or significantly different from the others, our estimates will not be as good.

Prime Numbers

A prime number is one that is evenly divisible only by itself and 1. (The number 1 (= 1 x 1) is not considered to be prime.) The prime numbers up to 100 are 2, 3, 5, 7, 11, 13, 17, 19, 23, 29, 31, 37, 41, 43, 47, 53, 59, 61, 67, 71, 73, 79, 83, 89, and 97. Note that 2 is the only even prime; also note that beyond 5, primes end only in 1, 3, 7, or 9.

SOLVING PICTOGRAM MESSAGES

The sample message from Figure 9.12: Read it in rows, from left to right, top row to bottom row.

```
11100000101010000000011100000100000001000000011000100000000000010
00001000000000000000000000000001000000000000000010000000000000000
00000000000010000010001000000010000000001110001000000000010000001
00000000000010000000000000000010000000010000000000000000000100000
00010000110000000001000000000011100001000000000010001111000000
01010000000100110000101100000000000000000001010010000000000000
00101110001100000000000010010001100000000000001001011000000010000
00010000110100001000101010000001111000010000000000000010001000011
0000000000001001100000100000000110000000000000100001001000000110
00000100111001000000000000100111100001000011010011001111110010000
00000111101111010010001111110111111000011000000000111101110100
10001000100110001111100100000000000000101100001000111110000000100
11100100000000000000010000100100101001000000000000000011000000011111
1111111100000001011
```

Step 1: Count the bits – the number of 1s and 0s. You should find 851.

Step 2: Factor the total number of bits (851) into two numbers. There should be only *one* solution, and both factors should be *prime numbers*. 851 = 23 x 37

Step 3: Lay out a rectangle on graph paper, with dimensions 23 x 37.

Step 4: Fill in the rectangle's grid using the bits from the message. (It is usually easier to see the picture if characters other than 1 and 0 are used, as shown here.)

Different pictures will be derived depending on whether filling is done by *rows* or by *columns*; Pictures A and B on the next page show the two principal options.

Picture A was filled row by row; Picture B was filled column by column. Arrows indicate the direction of fill of the first row (or column) of the message.

Step 5: If the filling is done correctly, the bits should *exactly* fill the rectangle, with no bits left over and no empty spaces at the end. Mistakes in filling should be corrected *now* before going on.

Step 6: *One* of these two pictures should appear to make some sense; the other should not. Choose the one that seems to be showing you something.

Sample pictogram array options:

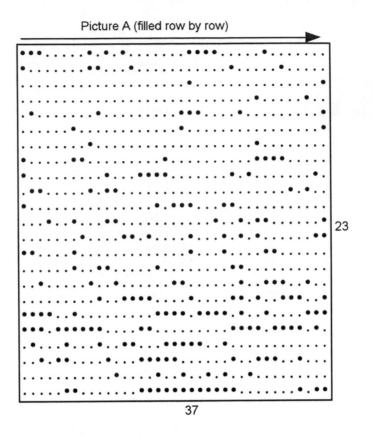

Picture A (filled row by row)

23

37

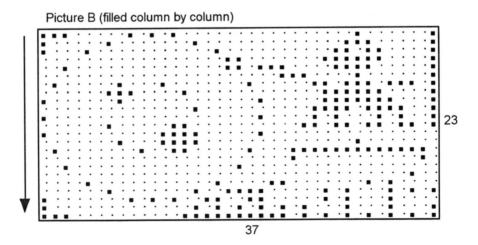

Picture B (filled column by column)

23

37

For this message, Picture B is obviously the one with which to work.

Step 7: Interpret the information contained in the pictogram. (Note: Aliens may have different conventions about left, right, up, and down.) Interpretation of Picture B is shown.

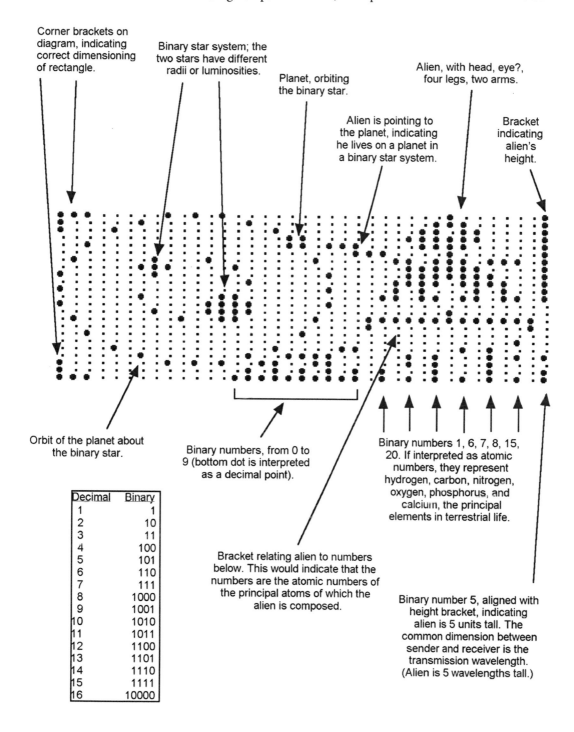

Corner brackets on diagram, indicating correct dimensioning of rectangle.

Binary star system; the two stars have different radii or luminosities.

Planet, orbiting the binary star.

Alien is pointing to the planet, indicating he lives on a planet in a binary star system.

Alien, with head, eye?, four legs, two arms.

Bracket indicating alien's height.

Orbit of the planet about the binary star.

Binary numbers, from 0 to 9 (bottom dot is interpreted as a decimal point).

Binary numbers 1, 6, 7, 8, 15, 20. If interpreted as atomic numbers, they represent hydrogen, carbon, nitrogen, oxygen, phosphorus, and calcium, the principal elements in terrestrial life.

Bracket relating alien to numbers below. This would indicate that the numbers are the atomic numbers of the principal atoms of which the alien is composed.

Binary number 5, aligned with height bracket, indicating alien is 5 units tall. The common dimension between sender and receiver is the transmission wavelength. (Alien is 5 wavelengths tall.)

Decimal	Binary
1	1
2	10
3	11
4	100
5	101
6	110
7	111
8	1000
9	1001
10	1010
11	1011
12	1100
13	1101
14	1110
15	1111
16	10000

PLANETARY DATA FOR THE SOLAR SYSTEM

Planet (*Dwarf Planet)	Radius (km)	Mass (Earth mass)	Density (g/cc)	Sidereal Rotation Period	Obliquity (degrees)	Average Distance from Sun (AU)	Orbital Period
Mercury	2439	0.055	5.43	58.65 d	0.0	0.387	87.97 d
Venus	6052	0.815	5.24	243 d	177.3	0.723	224.7 d
Earth	6378	1.000	5.52	23.93 h	23.45	1.000	365.256 d
Mars	3393	0.107	3.94	24.62 h	25.2	1.524	686.99 d
Jupiter	71,400	318	1.33	9.92 h	3.1	5.203	11.9 y
Saturn	60,000	95.2	0.70	10.5 h	26.7	9.555	29.5 y
Uranus	25,400	14.5	1.30	17.24 h	97.9	19.22	84.0 y
Neptune	24,300	17.2	1.76	16.05 h	29.6	30.11	165 y
Pluto*	1122	0.0025	2	6.39 d	118	39.44	248 y

REFERENCES

Pierce, James N. 2000. *Elementary astronomy.* Torrance, CA: Good Apple.

DRAKE EQUATION WORKSHEET

The Drake Equation: $N = N_* f_s N_p f_e f_l f_i f_c f_t$

N = the number of technical civilizations in the Galaxy at present

Factors	My Guess
N_* = the number of stars in our Galaxy	$N_* = $ _____
f_s = the fraction of all stars that are similar to the Sun	$f_s = $ _____
N_p = the average number of planets per Sun-like star	$N_p = $ _____
f_e = the fraction of planets (about Sun-like stars) that are similar to the Earth	$f_e = $ _____
f_l = the fraction of such planets on which life has actually developed	$f_l = $ _____
f_i = the fraction of these planets on which intelligent life has evolved	$f_i = $ _____
f_c = the fraction of these planets on which a technical civilization has developed	$f_c = $ _____

f_t = the fraction of time since the development of a planet's initial technical civilization that a technical civilization has existed on the planet; $f_t = L / t$ where L is the lifetime of an average technical civilization, and t is the average time elapsed since the inception of a technical civilization on each planet – estimated at about 3 billion years.

$f_t = L / (3\ \text{billion yrs})$ $\qquad\qquad\qquad$ L = _____ yrs

N = the number of Earth-like planets (orbiting Sun-like stars in the Galaxy) on which a technical civilization currently exists, which equals the number of technical civilizations in the Galaxy at present:

N_* x $\quad$ f_s x $\quad$ N_p x $\quad$ f_e x $\quad$ f_l x $\quad$ f_i x $\quad$ f_c x $\quad$ L / t $\qquad\quad$ = $\quad$ N

___ x $\quad$ ___ x $\quad$ ___ x $\quad$ ___ x $\quad$ ___ x $\quad$ ___ x $\quad$ ___ x $\quad$ ___ / (3 billion) = _____

References

Allen, C.W. 1973. *Astrophysical quantities*. 3rd ed. London: The Athlone Press.

Basri, Gibor. 2005. A decade of brown dwarfs. *Sky & Telescope* 109(5): 34-40.

Baugher, Joseph F. 1985. *On civilized stars: The search for intelligent life in outer space*. Englewood Cliffs, NJ: Prentice-Hall, Inc.

Bell, Jim. 2005. In search of Martian seas. *Sky & Telescope* 109(3): 40-47.

Breuer, Reinhard A. 1982. *Contact with the stars*. San Francisco: W.H. Freeman.

Carroll, Bradley W., and Dale A. Ostlie. 1996. *An introduction to modern astrophysics*. New York: Addison-Wesley.

Cox, Arthur N. ed. 2000. *Allen's astrophysical quantities*. 4th ed. New York: Springer-Verlag.

Darwin, Charles. 1859. *The origin of species by means of natural selection, or the preservation of favored races in the struggle for life*. New York: Modern Library.

Eldredge, Niles, and Stephen J. Gould. 1972. "Punctuated equilibrium": An alternative to phyletic gradualism. In *Models in paleobiology*, ed. T. J. M. Schopf, 82-115. San Francisco: Freeman, Cooper and Co.

Emsley, John. 1999. *The elements*. 3rd ed. New York: Oxford University Press.

Hart, Michael H. 1978. The evolution of the atmosphere of the Earth. *Icarus* 33: 23-39.

Hart, Michael H. 1979. Habitable zones about main sequence stars. *Icarus* 37: 351-57.

Huang, Su-Shu. 1975. Life outside the solar system. In *New frontiers in astronomy*, ed. Owen Gingerich, 104-112. San Francisco: W. H. Freeman. Originally published in *Scientific American* April 1960.

Kasting, James F. 1997. Habitable zones around low mass stars and the search for extraterrestrial life. *Origins of Life and Evolution of the Biosphere*. 27: 291-307.

Kasting, James F., Daniel P. Whitmire, and Ray T. Reynolds. 1993. Habitable zones around main sequence stars. *Icarus* 101: 108-128.

Lang, Kenneth R. 1980. *Astrophysical formulae.* 2nd ed. New York: Springer-Verlag.

Lide, David R. ed. 1998. *CRC handbook of chemistry and physics.* 79th ed. New York: CRC Press.

Lineweaver, Charles H., Yeshe Fenner, and Brad K. Gibson. 2004. The galactic habitable zone and the age distribution of complex life in the Milky Way. *Science* 303: 59-62.

Moore, Walter J. 1972. *Physical chemistry.* 4th ed. Englewood Cliffs, NJ: Prentice-Hall.

Pierce, James N. 2000. *Elementary astronomy.* Torrance, CA: Good Apple.

Powell, James L. 1998. *Night comes to the cretaceous: Comets, craters, controversy, and the last days of the dinosaurs.* San Diego: Harcourt Brace.

Stull, Daniel R., Edgar F. Westrum Jr., and Gerard C. Sinke. 1969. *The chemical thermodynamics of organic compounds.* New York: Wiley.

Reading List

There are a great many books addressing various aspects of the topic of extraterrestrial life. This list is intended to provide the reader with a few starting points for further investigation.

Friedman, Stanton T. 2008. *Flying saucers and science: A scientist investigates the mysteries of UFOs.* Franklin Lakes, NJ: New Page Books.

Behrendt, Kenneth W. 2007. *Secrets of UFO technology.* Bloomington, IN: AuthorHouse.

Bell, Kelly D. 2007. *Visitors: A new look at UFOs.* Lincoln, NE: iUniverse, Inc.

Bennett, Jeffrey, and Seth Shostak. 2007. *Life in the universe.* 2nd ed. San Francisco: Addison Wesley.

Brown, Bridget. 2007. *They know us better than we know ourselves: The history and politics of alien abduction.* New York: NYU Press.

Carey, Thomas J., and Donald R. Schmitt. 2007. *Witness to Roswell: Unmasking the 60-year cover-up.* Franklin Lakes, NJ: New Page Books.

Chester, Keith. 2007. *Strange company: Military encounters with UFOs in World War II.* San Antonio: Anomalist Books.

Friedman, Stanton T., and Kathleen Marden. 2007. *Captured! The Betty and Barney Hill UFO experience: The true story of the world's first documented alien abduction.* Franklin Lakes, NJ: New Page Books.

Genta, Giancarlo. 2007. *Lonely minds in the universe.* New York: Springer.

Good, Timothy. 2007. *Need to know: UFOs, the military, and intelligence.* New York: Pegasus Books.

Reece, Gregory L. 2007. *UFO religion: Inside flying saucer cults and culture.* New York: I. B. Tauris.

Sullivan, Woodruff T., and John Baross, eds. 2007. *Planets and life: The emerging science of astrobiology.* New York: Cambridge University Press.

Tumminia, Diana G. 2007. *Alien worlds: Social and religious dimensions of extraterrestrial contact.* Syracuse, NY: Syracuse University Press.

Basalla, George. 2006. *Civilized life in the universe.* New York: Oxford University Press.

Engdahl, Sylvia, ed. 2006. *Extraterrestrial life.* Detroit: Greenhaven Press.

Julien, Eric. 2006. *The science of extraterrestrials: UFOs explained at last.* Fort Oglethorpe, GA: Allies Publishing, Inc.

Michaud, Michael A. G. 2006. *Contact with alien civilizations: Our hopes and fears about encountering extraterrestrials.* New York: Springer.

Battaglia, Debbora, ed. 2005. *E.T. culture: Anthropology in outerspaces.* Durham, NC: Duke University Press.

Clancy, Paul, André Brack, and Gerda Horneck. 2005. *Looking for life, searching the solar system.* Cambridge: Cambridge University Press.

Clancy, Susan A. 2005. *Abducted: How people come to believe they were kidnapped by aliens.* Cambridge, MA: Harvard University Press.

Ward, Peter D. 2005. *Life as we do not know it: The NASA search for (and synthesis of) alien life.* New York: Viking.

Bova, Ben. 2004. *Faint echoes, distant stars: The science and politics of finding life beyond Earth.* New York: William Morrow.

Ehrenfreund, P., W. M. Irvine, T. Owen, L. Becker, J. Blank, J. R. Brucato, L. Colangeli, S. Derenne, A. Dutrey, D. Despois, A. Lazcano, and F. Robert, eds. 2004. *Astrobiology: Future perspectives.* New York: Springer.

Gilmour, Iain, and Mark A. Sephton, eds. 2004. *An introduction to astrobiology.* Cambridge: Cambridge University Press.

Lunine, Jonathan. 2004. *Astrobiology: A multi-disciplinary approach.* San Francisco: Benjamin Cummings.

Schulze-Makuch, Dirk, and Louis N. Irwin. 2004. *Life in the universe: Expectations and constraints.* New York: Springer.

Bennett, Jeffrey, Seth Shostak, and Bruce Jakosky. 2003. *Life in the universe.* San Francisco: Addison Wesley.

Burger, William C. 2003. *Perfect planet, clever species: How unique are we?* Amherst, NY: Prometheus Books.

Clark, Jerome. 2003. *Strange skies: Pilot encounters with UFOs.* New York: Citadel Press.

Cockell, Charles. 2003. *Impossible extinction: Natural catastrophes and the supremacy of the microbial world.* New York: Cambridge University Press.

Grinspoon, David H. 2003. *Lonely planets: The natural philosophy of alien life.* New York: ECCO.

Mayor, Michel, and Pierre-Yves Frei. 2003. *New worlds in the cosmos: The discovery of exoplanets.* New York: Cambridge University Press.

Moffitt, John F. 2003. *Picturing extraterrestrials: Alien images in modern culture.* Amherst, NY: Prometheus Press.

O'Neill, Terry, ed. 2003. *UFOs: Fact or fiction.* San Diego: Greenhaven Press.

Partridge, C. 2003. *UFO religions.* Boca Raton, FL: Routledge.

Shostak, Seth, and Alex Barnett. 2003. *Cosmic company: The search for life in the universe.* New York: Cambridge University Press.

Ulmschneider, Peter. 2003. *Intelligent life in the universe*. New York: Springer-Verlag.

Williams, Mary E., ed. 2003. *Paranormal phenomena*. San Diego: Greenhaven Press.

Cohen, Jack, and Ian Stewart. 2002. *What does a Martian look like? The science of extraterrestrial life*. Hoboken, NJ: J. Wiley.

Darling, David J. 2002. *The maverick science of astrobiology*. New York: Perseus Books Group.

Dorminey, Bruce. 2002. *Distant wanderers: The search for planets beyond the solar system*. New York: Copernicus.

Ekers, R. D., D. Kent Cullers, and John Billingham. 2002. *SETI 2020: A roadmap for the search for extraterrestrial intelligence*. Mountain View, CA: SETI Press.

Ward, Peter D., and Donald Brownlee. 2002. *The life and death of planet Earth: How the new science of astrobiology charts the ultimate fate of our world*. New York: Times Books.

Webb, Stephen. 2002. *If the universe is teeming with aliens – where is everybody?* New York: Copernicus Books.

Alschuler, William R. 2001. *The science of UFOs*. New York: St. Martin's Press.

Denzler, Brenda. 2001. *The lure of the edge: Scientific passions, religious beliefs, and the pursuit of UFOs*. Berkeley: University of California Press.

Goldsmith, Donald, and Tobias Owen. 2001. *The search for life in the universe*. 3rd ed. Sausalito, CA: University Science Books.

McConnell, Brian. 2001. *Beyond contact: A guide to SETI and communicating with alien civilizations*. Sebastopol, CA: O'Reilly.

Pflock, Karl T. 2001. *Roswell: Inconvenient facts and the will to believe*. Amherst, NY: Prometheus Books.

Roleff, Tamara L., ed. 2001. *Extraterrestrial life*. San Diego: Greenhaven Press.

Baker, Alan. 2000. *The encyclopedia of alien encounters*. New York: Facts on File.

Bragg, Desmond, and Paul Joslin. 2000. *Science meets the UFO enigma*. Huntington, NY: Kroschka Books.

Clark, Jerome. 2000. *Extraordinary encounters: An encyclopedia of extraterrestrials and otherworldly beings*. Santa Barbara, CA: ABC-CLIO.

Clark, Stuart G. 2000. *Life on other worlds and how to find it*. New York: Springer-Praxis.

Darling, David J. 2000. *The extraterrestrial encyclopedia: An alphabetical reference to all life in the universe*. New York: Three Rivers Press.

Jacobs, David M., ed. 2000. *UFOs and abductions: Challenging the borders of knowledge*. Lawrence, KS: University Press of Kansas.

Koerner, David, and Simon LeVay. 2000. *Here be dragons: The scientific quest for extraterrestrial life*. Oxford, NY: Oxford University Press.

Lewis, James R. 2000. *UFOs and popular culture: An encyclopedia of contemporary myth.* Santa Barbara, CA: ABC-CLIO.

Smith, Toby. 2000. *Little gray men: Roswell and the rise of a popular culture.* Albuquerque, NM: University of New Mexico Press.

Tough, Allen. 2000. *When SETI succeeds: The impact of high-information contact.* Bellevue, WA: Foundation For the Future.

Ward, Peter D., and Donald Brownlee. 2000. *Rare Earth: Why complex life is uncommon in the universe.* New York: Copernicus.

Achenbach, Joel. 1999. *Captured by aliens: The search for life and truth in a very large universe.* New York: Simon & Schuster.

Chambers, Paul. 1999. *Life on Mars: The complete story.* London: Blandford.

Clark, Andrew J.H., and David H. Clark. 1999. *Aliens: Can we make contact with extraterrestrial intelligence.* New York: Fromm International.

Davies, Paul. 1999. *The fifth miracle: The search for origin and meaning of life.* New York: Simon & Schuster.

Dudley, William, ed. 1999. *UFOs.* San Diego: Greenhaven Press.

Gold, Thomas. 1999. *The deep hot biosphere.* New York: Copernicus.

Greer, Steven M. 1999. *Extraterrestrial contact: The evidence and implications.* Afton, VA: Crossing Point Inc.

Pope, Nick. 1999. *Open skies, closed minds: For the first time a government UFO expert speaks out.* Woodstock, NY: Overlook Press.

Shapiro, Robert. 1999. *Planetary dreams: The quest to discover life beyond Earth.* New York: J. Wiley.

Sturrock, Peter A. 1999. *The UFO enigma: A new review of the physical evidence.* New York: Warner Books.

Taylor, Michael R. 1999. *Dark life: Martian nanobacteria, rock-eating cave bugs, and other extreme organisms of inner Earth and outer space.* New York: Scribner.

Walter, Malcolm. 1999. *The search for life on Mars.* Cambridge, MA: Perseus.

Aczel, Amir D. 1998. *Probability 1: Why there must be intelligent life in the universe.* New York: Harcourt Brace.

Bartholomew, Robert E., and George S. Howard. 1998. *UFOs and alien contact: Two centuries of mystery.* Amherst, NY: Prometheus Books.

Clark, Jerome. 1998. *The UFO encyclopedia: The phenomenon from the beginning.* 2nd ed. Detroit: Omnigraphics.

Däniken, Erich von. 1998. *Arrival of the gods: Revealing the alien landing sites of Nazca.* n.p.: Diane Pub. Co.

Dean, Jodi. 1998. *Aliens in America: Conspiracy cultures from outerspace to cyberspace.* Ithaca, NY: Cornell University Press.

Dick, Steven J. 1998. *Life on other worlds: The 20th century extraterrestrial life debate.* Cambridge: Cambridge University Press.

Fisher, David E., and Marshall J. Fisher. 1998. *Strangers in the night: A brief history of life on other worlds.* Washington, DC: Counterpoint.

Fitzgerald, Randall. 1998. *Cosmic test tube: Extraterrestrial contact, theories & evidence.* Los Angeles: Moon Lake Media.

Hesemann, Michael. 1998. *UFOs: The secret history.* New York: Marlowe & Company.

Hesemann, Michael, and Philip Mantle. 1998. *Beyond Roswell: The alien autopsy film, Area 51, & the U.S. Government coverup of UFOs.* New York: Marlowe & Company.

Jakosky, Bruce. 1998. *The search for life on other planets.* Cambridge, UK: Cambridge University Press.

Lemonick, Michael D. 1998. *Other worlds: The search for life in the universe.* New York: Simon & Schuster.

Lewis, John S. 1998. *Worlds without end: The exploration of planets known and unknown.* Reading, MA: Perseus Books.

Matheson, Terry. 1998. *Alien abductions: Creating a modern phenomenon.* Amherst, NY: Prometheus Books.

Parker, Barry R. 1998. *Alien life: The search for extraterrestrials and beyond.* New York: Plenum Trade.

Pickover, Clifford A. 1998. *The science of aliens.* New York: Basic Books.

Powell, James L. 1998. *Night comes to the cretaceous: Comets, craters, controversy, and the last days of the dinosaurs.* San Diego: Harcourt Brace.

Shostak, Seth. 1998. *Sharing the universe: Perspectives on extraterrestrial life.* Berkeley, CA: Berkeley Hills Books.

Wilson, Colin. 1998. *Alien dawn: An investigation into the contact experience.* New York: Fromm International.

Däniken, Erich von. 1997. *The return of the gods: Evidence of extraterrestrial visitations.* Shaftesbury, Dorset, UK: Element.

Devereux, Paul, and Peter Brookesmith. 1997. *UFOs and ufology: The first 50 years.* New York: Facts on File.

Emmons, Charles F. 1997. *At the threshold: UFOs, science and the New Age.* Mill Spring, NC: Wild Flower Press.

Fawcett, Bill, ed. 1997. *Making contact: A serious handbook for locating and communicating with extraterrestrials.* New York: Avon Books.

Fortey, Richard. 1997. *Life: A natural history of the first four billion years of life on Earth.* New York: Vintage Books.

Frazier, Kendrick, Barry Karr, and Joe Nickell, eds. 1997. *The UFO invasion: The Roswell incident, alien abductions, and government cover-ups.* Amherst, NY: Prometheus Books.

Goldsmith, Donald. 1997. *The hunt for life on Mars.* New York: Dutton.

Halpern, Paul. 1997. *The quest for alien planets: Exploring worlds outside the solar system.* New York: Plenum Trade.

Harrison, Albert A. 1997. *After contact: The human response to extraterrestrial life.* New York: Plenum Trade.

Korff, Kal K. 1997. *Roswell UFO crash: What they don't want you to know.* Amherst, NY: Prometheus Books.

Marrs, Jim. 1997. *Alien agenda: Investigating the extraterrestrial presence among us.* New York: HarperCollins.

McAndrew, James. 1997. *The Roswell report: Case closed.* Washington, DC: Headquarters United States Air Force.

McSween, Harry Y. 1997. *Fanfare for Earth: The origin of our planet and life.* New York: St. Martin's Press.

Randles, Jenny. 1997. *Alien contact: The first fifty years.* New York: Sterling Publishing.

Sellier, Charles E. 1997. *UFO.* Chicago: Contemporary Books.

Ashpole, Edward. 1996. *The UFO phenomena: A scientific look at the evidence for extraterrestrial contacts.* London: Headline.

Brown, Courtney. 1996. *Cosmic voyage: A scientific discovery of extraterrestrials visiting Earth.* New York: Dutton.

Craft, Michael. 1996. *Alien impact: A comprehensive look at the evidence of human-alien contact.* New York: St. Martin's Press.

Dick, Steven J. 1996. *The biological universe: The 20th century extraterrestrial life debate and the limits of science.* Cambridge: Cambridge University Press.

Bryan, Courtlandt D. B. 1995. *Close encounters of the fourth kind: Alien abduction, UFOs, and the conference at M.I.T.* New York: Knopf.

Davies, Paul. 1995. *Are we alone?: Philosophical implications of the discovery of extraterrestrial life.* New York: Basic Books.

Hart, Michael H., and Ben Zuckerman, eds. 1995. *Extraterrestrials: Where are they?* Cambridge: Cambridge University Press.

Heidmann, Jean. 1995. *Extraterrestrial intelligence.* Cambridge: Cambridge University Press.

Hesemann, Michael. 1995. *Cosmic connection: Worldwide crop formations and ET contacts.* Bath, UK: Gateway.

Lewis, James R., ed. 1995. *The gods have landed: New religions from other worlds.* Albany: State University of New York Press.

Weaver, Richard L., and James McAndrew. 1995. *The Roswell report: Fact versus fiction in the New Mexico desert.* Washington, DC: Headquarters United States Air Force.

Mack, John E. 1994. *Abduction: Human encounters with aliens.* New York: Scribner's.

Peebles, Curtis. 1994. *Watch the skies! A chronicle of the flying saucer myth.* Washington: Smithsonian Institution Press.

Good, Timothy. 1993. *Alien contact: Top-secret UFO files revealed.* New York: W. Morrow.

Sullivan, Walter. 1993. *We are not alone: The continuing search for extraterrestrial intelligence.* New York: Dutton.

Drake, Frank, and Dava Sobel. 1992. *Is anyone out there? The scientific search for ETI.* New York: Delacorte Press.

Editors of Time-Life Books. 1992. *Mysteries of the unknown: Alien encounters.* Alexandria, VA: Time-Life Books.

Heidmann, Jean. 1992. *Life in the universe.* New York: McGraw-Hill.

Jacobs, David M. 1992. *Secret life: Firsthand accounts of UFO abductions.* New York: Simon & Schuster.

Randles, Jenny. 1992. *UFOs and how to see them.* New York: Sterling Pub. Co.

Ring, Kenneth. 1992. *The Omega Project: Near-death experiences, UFO encounters, and mind at large.* New York: William Morrow.

Angelo Jr., Joseph A. 1991. *The extraterrestrial encyclopedia: Our search for life in outer space.* New York: Facts on File.

Davoust, Emmanuel. 1991. *The cosmic water hole.* Cambridge, MA: MIT Press.

Randle, Kevin D., and Donald R. Schmitt. 1991. *UFO crash at Roswell.* New York: Avon Books.

Blum, Howard. 1990. *Out there: The government's secret quest for extraterrestrials.* New York: Simon & Schuster.

Bova, Ben, and Byron Preiss, eds. 1990. *First contact: The search for extraterrestrial intelligence.* New York: Penguin Books.

Swift, David W. 1990. *SETI pioneers: Scientists talk about their search for extraterrestrial intelligence.* Tucson: University of Arizona Press.

White, Frank. 1990. *The SETI factor: How the search for extraterrestrial intelligence is changing our view of the universe and ourselves.* New York: Walker.

Good, Timothy. 1989. *Above top secret: The worldwide UFO cover-up.* New York: Quill.

Klass, Philip J. 1989. *UFO abductions: A dangerous game.* Buffalo: Prometheus Books.

Editors of Time-Life Books. 1987. *Mysteries of the unknown: The UFO phenomenon.* Alexandria, VA: Time-Life Books.

Jackson, Francis, and Patrick Moore. 1987. *Life in the universe.* New York: Norton.

Kutter, G. Siegfried. 1987. *The universe and life: Origins and evolution.* Boston: Jones and Bartlett.

McDonough, Thomas R. 1987. *The search for extraterrestrial intelligence.* New York: John Wiley & Sons.

Crowe, Michael J. 1986. *The extraterrestrial life debate, 1750-1900: The idea of a plurality of worlds from Kant to Lowell.* Cambridge: Cambridge University Press.

Horowitz, Norman H. 1986. *To utopia and back: The search for life in the solar system.* New York: W.H. Freeman.

Baugher, Joseph F. 1985. *On civilized stars: The search for intelligent life in outer space.* Englewood Cliffs, NJ: Prentice-Hall, Inc.

Regis Jr., Edward, ed. 1985. *Extraterrestrials: Science and alien intelligence.* Cambridge: Cambridge University Press.

Elders, Lee J., Brit Nilsson-Elders, and Thomas K. Welch. 1983. *UFO: Contact from the Pleiades.* Phoenix: Genesis III Productions.

Klass, Philip J. 1983. *UFOs: The public deceived.* Buffalo: Prometheus Books.

Breuer, Reinhard A. 1982. *Contact with the stars.* San Francisco: W.H. Freeman.

Dick, Steven J. 1982. *Plurality of worlds: The origins of the extraterrestrial life debate from Democritus to Kant.* Cambridge: Cambridge University Press.

Oberg, James E. 1982. *UFOs and outer space mysteries: A sympathetic skeptic's report.* Norfolk, VA: Donning.

Billingham, John, ed. 1981. *Life in the universe.* Cambridge, MA: MIT Press.

Gribbin, John. 1981. *Genesis: The origins of man and the universe.* New York: Delacorte Press.

Rood, Robert T., and James S. Trefil. 1981. *Are we alone? The possibility of extraterrestrial civilizations.* New York: Scribner.

Sheaffer, Robert. 1981. *The UFO verdict: Examining the evidence.* Buffalo: Prometheus Books.

Berlitz, Charles, and William Moore. 1980. *The Roswell incident.* New York: Grosset & Dunlap.

Cooper, Henry S. F. 1980. *The search for life on Mars: Evolution of an idea.* New York: Holt, Rinehart and Winston.

Feinberg, Gerald, and Robert Shapiro. 1980. *Life beyond Earth: The intelligent Earthling's guide to life in the universe.* New York: Morrow.

Sagan, Carl. 1980. *Cosmos.* New York: Random House.

Asimov, Isaac. 1979. *Extraterrestrial civilizations.* New York: Crown Publishers.

Hoyle, Fred, and Chandra Wickramasinghe. 1979. *Diseases from space.* New York: Harper & Row.

Hoyle, Fred, and Chandra Wickramasinghe. 1978. *Lifecloud, the origin of life in the universe.* New York: Harper & Row.

Jung, C. G. 1978. *Flying saucers: A modern myth of things seen in the skies.* Princeton: Princeton University Press.

Ridpath, Ian. 1978. *Messages from the stars: Communication and contact with extraterrestrial life.* New York: Harper & Row.

Sagan, Carl, F. D. Drake, Ann Druyan, Timothy Ferris, Jon Lomberg, and Linda Salzman. 1978. *Murmurs of Earth: The Voyager interstellar record.* New York: Random House.

Steiger, Brad. 1978. *Worlds before our own.* New York: Berkley Pub. Corp.

Macvey, John W. 1977. *Interstellar travel: Past, present and future.* Briarcliff Manor, NY: Stein and Day.

Menzel, Donald H., and Ernest H. Taves. 1977. *The UFO enigma: The definitive explanation of the UFO phenomenon.* Garden City, NY: Doubleday.

Ridpath, Ian. 1976. *Worlds beyond: A report on the search for life in space.* New York: Harper & Row.

Bracewell, Ronald N. 1975. *The galactic club: Intelligent life in outer space.* San Francisco: W. H. Freeman.

Jacobs, David M. 1975. *The UFO controversy in America.* Bloomington, IN: Indiana University Press.

Landsburg , Alan. 1975. *Outer space connection.* New York: Bantam Books.

Maruyama, Magoroh, and Arthur Harkins, eds. 1975. *Cultures beyond the Earth.* New York: Vintage Books.

Ponnamperuma, Cyril, and A. G. W. Cameron, eds. 1974. *Interstellar communications: Scientific perspectives.* Boston: Houghton Mifflin.

Berendzen, Richard, ed. 1973. *Life beyond Earth & the mind of man; A symposium.* Washington: NASA.

Macvey, John W. 1973. *Whispers from space.* New York: Macmillan.

Sagan, Carl. 1973. *The cosmic connection: An extraterrestrial perspective.* Garden City: Anchor Press.

Däniken, Erich von. 1972. *The gold of the gods.* New York: Putnam.

Ponnamperuma, Cyril, ed. 1972. *Exobiology.* Amsterdam: North-Holland Pub. Co.

Sagan, Carl, and Thornton Page, eds. 1972. *UFO's–A scientific debate.* Ithaca, NY: Cornell University Press.

Däniken, Erich von. 1971. *Chariots of the gods? Unsolved mysteries of the past.* New York: Putnam.

Flammonde, Paris. 1971. *The age of flying saucers; Notes on a projected history of unidentified flying objects.* New York: Hawthorne Books.

Däniken, Erich von. 1970. *Gods from outer space: Return to the stars, or evidence for the impossible.* New York: Putnam.

Dole, Stephen H. 1970. *Habitable planets for man.* 2nd ed. New York: American Elsevier Pub. Co.

Leonard, R. Cedric. 1969. *Flying saucers, ancient writings and the Bible.* New York: Exposition Press.

Vallee, Jacques. 1969. *Passport to Magonia: From folklore to flying saucers.* Chicago: H. Regnery Co.

Young, Richard S. 1969. *Life beyond Earth.* Morristown, NJ: Silver Burdett Co.

Edwards, Frank. 1967. *Flying saucers, here and now!* New York: L. Stuart.

Cade, Cecil Maxwell. 1966. *Other worlds than ours.* New York: Taplinger Pub. Co.

Edwards, Frank. 1966. *Flying saucers – Serious business.* New York: Bantam Books.

MacGowan, Roger A., and Frederick I. Ordway. 1966. *Intelligence in the universe.* Englewood Cliffs, NJ: Prentice-Hall.

Shklovskii, I.S., and Carl Sagan. 1966. *Intelligent life in the universe.* New York: Dell Publishing.

Vallee, Jacques. 1965. *Anatomy of a phenomenon.* Chicago: H. Regnery Co.

Berrill, Norman J. 1964. *Worlds without end; A reflection on planets, life, and time.* New York: Macmillan.

Anderson, Poul. 1963. *Is there life on other worlds?* New York: Crowell-Collier Press.

Cameron, Alastair G. W., ed. 1963. *Interstellar communication.* New York: W. A. Benjamin.

Firsoff, Valdemar A. 1963. *Life beyond the Earth; A study in exobiology.* New York: Basic Books.

Drake, Frank D. 1962. *Intelligent life in space.* New York: Macmillan.

Ovenden, Michael W. 1962. *Life in the universe: A scientific discussion.* Garden City, NY: Anchor Books.

Oparin, Aleksandr I., and V. Fesenkov. 1961. *Life in the universe.* New York: Twayne Publishers.

Menzel, Donald H. 1953. *Flying saucers.* Cambridge, MA: Harvard University Press.

Jones, Harold S., Sir. 1952. *Life on other worlds.* 2nd ed. New York: Macmillan.

Heard, Gerald. 1951. *Is another world watching? The riddle of the flying saucers.* New York: Harper.

Glossary

A

abundance – a measure of the frequency with which a particular element occurs in a sample; may be expressed as a fraction of the whole or as a ratio to some reference element.

aerobic respiration – the process of obtaining energy by chemically combining organic molecules with atmospheric oxygen.

albedo – the fraction of incident radiation reflected by a body.

Alpha Centauri –the closest binary star to the Sun at a distance of 4.4 light-years, consisting of stars of spectral types G2 and K5. Its nearby neighbor, the fainter Proxima Centauri (spectral type M5), is slightly closer to the Sun (4.2 ly) and may be gravitationally bound to Alpha Centauri.

amino acid – an organic molecule containing a carbon atom bonded to both a carboxylic acid group (-COOH) and an amino group ($-NH_2$).

Andromeda Galaxy – the closest large galaxy to the Milky Way. Also known as M31, the Andromeda Galaxy is a spiral galaxy located about 2.9 million light-years from the Milky Way.

Ångstrom – a unit of length equal to one 100 millionth of a centimeter, the approximate diameter of an atom.

asteroid – (or minor planet) a small, usually rocky, body orbiting a star.

asteroid belt – the region of the solar system between the orbits of Mars and Jupiter where most of the asteroid orbits appear to be concentrated.

asteroid impact – collision of an asteroid with the Earth (or another planet or moon); such an event is most likely responsible for the extinction of the dinosaurs 65 million years ago.

astronomical unit (AU) – the average distance between the Earth and Sun.

asymmetric carbon – a carbon atom that is bonded to four different atoms or groups of atoms.

atmosphere – the gaseous envelope that lies above the surface of a planet or star.

atom – the smallest particle of an element that exhibits all of the element's chemical properties; an atom consists of a nucleus of protons and neutrons surrounded by a cloud of electrons.

atomic mass number – the number of protons and neutrons in an atom's nucleus – a measure of its approximate mass.

atomic number – the number of protons in an atom's nucleus.

atomic weight – a measure of the average mass of an atom of a particular element – an average made over the naturally occurring isotopes of the element, allowing for their different abundances.

autotroph – plant-like organism capable of using carbon dioxide as its sole carbon source.

$\mathcal{B}$

banded iron formations – sedimentary rock formations of alternating colors caused by the presence or absence of iron in its oxidized state; they are linked to variations in the abundance of atmospheric oxygen.

bandwidth – the range of frequencies over which a signal is transmitted or a detector is sensitive.

Big Bang – the start of the primeval fireball expansion that marked the beginning of the universe.

Big Bang Theory – the cosmological theory that holds that the universe began as a state of extremely high density and temperature and has expanded continuously since then.

binary code – a system of coding messages using only two different characters, such as 0 and 1.

binary star – two stars that orbit each other, bound together by their mutual gravitational attraction; most binary stars appear as single stars to the unaided eye.

black hole – an object with a gravitational field so intense that even light cannot escape from it. Black holes can be formed by the evolution of massive stars; super-massive black holes appear to exist at the centers of many galaxies.

brown dwarfs – objects intermediate in mass between planets and stars, in the general range of 13 to 80 Jupiter masses.

C

Cambrian explosion – a relatively brief period between about 570 and 530 million years ago when the lineages of most of the major divisions of the animal kingdom were established.

carbohydrates – a class of molecules – including sugars and starches – composed of carbon, hydrogen, and oxygen, the majority of which have empirical formulae of the form $C_n(H_2O)_n$.

carbon chauvinism – the attitude that extraterrestrial life will be based on carbon because terrestrial life is based on carbon.

chemistry – the study of the way atoms interact with each other.

closed universe – a Big Bang model of the universe in which the outrushing galaxies are slowed, halted, and ultimately pulled back together by gravity.

comet – a small, usually icy, body orbiting the Sun (and presumably other stars); as they approach the inner solar system, vaporization of some of the comet's frozen gases causes it to form a visible tail.

continental drift – the motion of the Earth's continental plates, caused by plate tectonics.

continuously habitable zone (CHZ)– the region around a star in which the star's radiation produces conditions on suitable planets allowing life to be supported continuously over the length of time required for intelligent life to evolve.

convection – a heat transfer mechanism in which warm fluid rises and cooler, denser fluid sinks.

core – the central region of a planet or star.

cosmological principle – on the large scale, the universe looks the same to all observers; there are no special positions in the universe.

cosmology – the study of the nature, origin, and evolution of the universe.

crop circles – geometric patterns formed overnight in fields of grain; occasionally attributed to aliens, their construction has been demonstrated by numerous teams of humans.

crust – the thin, outermost layer of a terrestrial planet, which defines the planet's surface.

cyanobacteria – blue-green algae; prokaryotes capable of photosynthesis.

$\mathcal{D}$

dark energy – an unknown entity that appears to be accelerating the rate at which the universe is expanding.

daughter isotope – a nucleus produced by radioactive decay of an unstable (parent) isotope.

deoxyribonucleic acid (DNA) – a nucleic acid produced from nucleotides based on the sugar D-2-deoxyribose; it forms the basis of the genetic code in terrestrial life.

differentiation – the process by which dense materials in a planetary interior sink toward the core, resulting in a variation in composition with radius.

disk – the flattened portion of a spiral galaxy; it contains the orbits of most of the stars.

Doppler shift – the shift in the frequency or wavelength of waves, caused by relative motion between the source and the observer.

double bond – a molecular bond between two atoms in which they each contribute two electrons to be shared.

Drake equation – a mathematical tool used to estimate the number of intelligent civilizations within the Milky Way; its factors include estimates of the probabilities of suitable stars, suitable planets, the emergence of life, the development of technical civilizations, etc.

dwarf planet – an object that orbits the Sun, is sufficiently massive to form itself into an essentially spherical shape, but has not cleared the neighborhood around its orbit, and is not a satellite (2006).

Dyson sphere – a sphere constructed around a star by an advanced civilization, to capture most or all of the energy emitted by the star.

ℰ

eccentricity – a measure of the shape of an elliptical orbit.

electric charge – a property of some elementary particles that causes them to exert electrostatic forces on each other; the two types of charges are designated (arbitrarily) as positive and negative.

electromagnetic radiation – traveling fluctuations in the electric and magnetic fields in space.

electromagnetic spectrum – all the different types of electromagnetic radiation as distinguished by their different wavelengths (or frequencies, or energies).

electron – the negatively charged elementary particle that resides in an atom's electron cloud; atoms form chemical bonds by sharing or exchanging electrons.

electron cloud – the region, surrounding the nucleus, in which the atom's electrons are located.

element – matter comprised of atoms that all have the same atomic number; about 100 different elements are known.

elliptical galaxy – a galaxy that displays an elliptical profile, without evidence of any disk, spiral arms, or nuclear bulge.

enantiomers – molecules that differ from each other only in the three-dimensional orientation of their bonds, such that they are mirror images of each other.

endothermic – energy absorbing; endothermic reactions absorb more energy from their surroundings than they produce.

energy – the capacity for doing work; the energy of an electromagnetic wave is proportional to its frequency.

eukaryote – the cell that makes up all of complex life on Earth; it normally contains organelles, generates energy by aerobic metabolism, and has a nucleus with DNA inside.

exoplanet – an extrasolar planet.

exothermic – energy releasing; exothermic reactions produce more energy than they absorb from their surroundings.

extrasolar planets – planets orbiting stars other than the Sun.

extraterrestrial – from beyond Earth.

F

fact – a well-established observation – one that can be made repeatedly.

fossil – the remains or traces of a life form, preserved in rock.

frequency – the number of oscillations that occur each second.

G

galactic cluster – see open cluster.

Galactic habitable zone – a ring-shaped zone (containing the Earth and Sun) around the center of the Galaxy that is suggested to be most likely to harbor planetary life; planetary systems farther out in the Galaxy may lack suffcient metals (due to low supernova rates) to form suitable terrestrial planets while stars and planets closer in are exposed to the hazards provided by frequent nearby supernovae.

galaxy – a vast collection of stars, their planetary systems, and nebulae, all bound together by gravity; galaxies typically contain billions of stars. When capitalized (e.g. the Galaxy), it refers to the Milky Way Galaxy.

gametes – special cells used in sexual reproduction (sperm and egg), each containing only half the DNA required for the organism.

giant – one of a group of stars on the right-center portion of the Hertzsprung-Russell diagram; they are generally larger and cooler than their main sequence predecessors (stars of medium mass).

globular cluster – a group of typically 100,000 to 1,000,000 stars, bound together by their mutual gravitational attraction; they are found in a spherical distribution about the center of a galaxy.

greenhouse effect – the process by which a planet absorbs radiation from its sun and then reradiates this energy at a wavelength that is absorbed by the planet's atmosphere, thus maintaining a higher surface temperature on the planet.

H

habitable zone – the region around a star in which the star's radiation produces conditions on suitable planets that allow them to support life.

half-life – the time required for one half of the atoms of a given radioactive isotope to decay.

halo – a sparsely populated, spherical distribution of stars surrounding a spiral galaxy.

heat capacity – the energy required to raise the temperature of one gram of a substance by one degree Celsius.

heat of vaporization – the energy required to vaporize one gram of a liquid.

Hertzsprung-Russell diagram (or **HR diagram**) – a graph of luminosity vs. temperature for the stars; on it, four groups of stars can be identified: main sequence, giants, supergiants, and white dwarfs.

heterotroph – animal-like organism capable of using organic matter as its sole carbon source.

Hubble law – most of the galaxies are moving away from us with speeds proportional to their distances; the farther away the galaxies are, the faster they are moving.

hypothesis – an explanation for a set of observations.

I

igneous rock – rock formed by the cooling of magma (molten rock).

ion – an atom or molecule that has lost or gained one or more electrons such that it carries a net charge.

irregular galaxy – a galaxy that shows no obvious structural features, such as a disk, nuclear bulge, or spiral arms.

isotopes – atoms with the same number of protons and different numbers of neutrons; a given element may exist as several different isotopes.

J

Jovian planets – large, low-density, gaseous (Jupiter-like) planets.

K

Kardashev civilizations – three different levels of technology proposed by the Soviet astronomer Nikolai S. Kardashev; Type I civilizations are similar to the one found on Earth (ours) in that they are able to utilize the power output of a planet; Type II civilizations can utilize the entire power output of a star, while Type III civilizations would control the power of a whole galaxy.

L

law – an extremely well-established (and usually very simple) theory.

length contraction – the foreshortening of a moving object in the direction of its motion as measured by an observer in a different reference frame.

Local Group – the cluster of galaxies in which the Milky Way Galaxy resides.

luminosity – the rate at which a star radiates energy.

M

magnetosphere – the region of space around a planet (or star) that is dominated by its magnetic field.

main sequence – a group of stars on the Hertzsprung-Russell diagram, ranging from low luminosity, low temperature stars to high luminosity, high temperature stars; the longest phase of a star's lifetime, during which hydrogen is converted to helium in the stellar core.

mantle – the region of a planetary interior that overlies the core.

mass – a measure of the amount of matter contained in a body.

mass extinction – an event that eliminates a significant fraction of the species on a planet.

meiosis – the process of cell reproduction in which two different cells (egg and sperm) combine, with each contributing half of the new cell's DNA, making this daughter cell different from either parent.

metals – (as used by astronomers) elements other than hydrogen and helium.

metamorphic rock – rock that has been altered by the effects of heat and pressure; the initial rock may be sedimentary, igneous, or metamorphic.

meteor – the flash of light observed when a meteoroid plunges through the atmosphere, heating the air to incandescence.

meteorite – a meteoroid that survives passage through the atmosphere and lands on the ground.

meteoroid – a relatively small rock in space in orbit about the Sun; its passage through a planetary atmosphere produces a meteor.

Milky Way – the faint band of light that circles the entire sky: our edge-on view of our galaxy; the spiral galaxy that contains the Sun and about 200 billion other stars.

mitosis – the process of cell reproduction in which the cell divides and its DNA creates a copy of itself, making the two resulting cells essentially identical.

molecule – two or more atoms bonded together; atoms accomplish this by shedding, stealing, or sharing electrons.

moon – a natural satellite of a planet.

N

natural satellite – a naturally occurring body that orbits another body.

natural selection – the process in which, within a given species, the individuals that are best adapted to their environment will be most likely to survive and reproduce; their traits will be selected by nature to be passed on to the following generations.

nebula – a cloud of interstellar gas and/or dust.

neutron – an elementary particle with no electric charge and a mass similar to that of the proton; found in the nucleus of the atom.

neutron star – a 'star' consisting of a sphere of degenerate neutrons; only a few kilometers in radius, it is formed by the collapse of the core of a massive star during a supernova event.

noise temperature – a temperature value used to measure the power of noise generated by electronic equipment and/or astrophysical sources; higher temperature indicates greater noise.

nuclear bulge – a spherical distribution of stars at the center of a spiral galaxy.

nuclear fission – a reaction in which an atomic nucleus splits apart to form smaller nuclei.

nuclear fusion – a reaction in which atomic nuclei combine to form a larger nucleus.

nucleosynthesis – the manufacture of different atomic nuclei through the processes of fusion, fission, and radioactive decay.

nucleotide – a molecular unit consisting of a sugar, a phosphate group, and a nitrogen base; these units combine to form the nucleic acids RNA and DNA.

nucleus – the central region of an atom where most of its mass is concentrated; it consists of protons and neutrons. Also, the central region of a galaxy.

number abundance – the abundance of a particular element, expressed in terms of the numbers of atoms of that element relative to either the total number of atoms in the sample or the number of atoms of some reference element.

O

obliquity – the angle between a planet's rotational axis and its orbital axis.

observation – a report of a measurement performed or a phenomenon that is witnessed.

Occam's razor – a guideline, attributed to William of Occam, which states that when choosing among multiple hypotheses, one should select the explanation that requires the fewest assumptions to be made; in short, the simpler the theory, the better.

open cluster – a group of 100 to 1000 stars, bound together by their mutual gravitational attraction; they form and orbit within the plane of a spiral galaxy.

open universe – a Big Bang model of the universe in which the outrushing galaxies continue to move away from each other over time.

organic chemistry – the study of organic molecules, their properties, their formation, and the reactions they undergo.

organic molecule – a molecule containing carbon-hydrogen bonds.

ozone layer – a region in the stratosphere where ozone (O_3) is abundant; these molecules absorb ultraviolet radiation, helping to protect terrestrial life from the Sun's harmful rays.

℘

parent isotope – a nucleus that is unstable to radioactive decay, which transforms it into a daughter isotope.

periodic table of the elements – a chart that presents the elements in order of increasing atomic number, grouped together by their common electronic structure.

photon – a 'particle' of electromagnetic radiation.

photosynthesis – the process by which plants combine carbon dioxide, water, and energy from sunlight to make sugar, releasing oxygen as a byproduct.

phyletic gradualism – the theory that new species form gradually over time by constant modification of existing species.

pictogram – a linear string of bits arranged in a two-dimensional array to form a picture.

planet – one of the bodies orbiting the Sun (or another star), typically more massive than a comet or an asteroid, but less massive than a brown dwarf; an object that orbits the Sun, is sufficiently massive to form itself into an essentially spherical shape, and has cleared the neighborhood around its orbit (2006).

planetary nebula – a luminous gaseous nebula formed by the non-explosive ejection of the outer layers of an evolving giant.

planetesimal – small bodies orbiting the Sun that represent the stage of accretion just before the formation of planets.

plate tectonics – the process by which a planet's crustal plates are carried slowly across the surface by convective currents in the mantle.

Population I – a population of stars and other objects characterized by a relatively high metal abundance.

Population II – a population of stars and other objects characterized by a relatively low metal abundance.

precession – the gradual shift in the orientation of the Earth's rotational axis with respect to the stars.

prime number – a number evenly divisible only by itself and 1 (1 is not considered prime).

primeval fireball – the high temperature, high density state that existed at the start of the universe.

principle of mediocrity – the assumption that there is nothing special about our Sun or solar system, that our planetary system is typical of those found throughout the Galaxy.

probability – the likelihood of the occurrence of an event.

Project Ozma – the first search for extraterrestrial radio signals, conducted by Frank Drake.

prokaryote – a primitive cell, containing no nucleus.

proper length – the length of an object measured by an observer at rest with respect to the object; the proper length is the greatest length measurable for a given dimension.

proper time – the duration of an event measured by an observer in the reference frame in which the event occurs; the proper time is the shortest time measurable for an event.

protein – large molecules consisting of long chains of amino acids linked together.

proton – the positively charged elementary particle that resides in the nucleus of an atom.

protostar – a forming star that is undergoing gravitational contraction prior to the ignition of stabilizing nuclear reactions in its core.

pulsar – a radio source pulsating with a very short period, attributed to a rapidly spinning neutron star.

punctuated equilibrium – the theory that species remain fairly constant over long periods of time, only changing – relatively rapidly – when new environmental conditions are imposed.

R

radioactive decay – the transformation of one element into another by spontaneous emission of a particle or photon by an unstable nucleus.

reducing atmosphere – an atmosphere rich in hydrogen and hydrogen compounds, such as methane and ammonia (the reduced forms of carbon and nitrogen).

reference frame – the set of coordinates used to measure the position, velocity, etc. of an object.

rest mass – the mass of an object measured by an observer at rest with respect to the object; the rest mass is the smallest mass measurable for an object.

ribonucleic acid (RNA) – a nucleic acid produced from nucleotides based on the sugar D-ribose.

runaway icebox – a situation occurring when a planet cools such that its water turns to ice, a more efficient reflector of sunlight that promotes further cooling.

S

satellite – a body that orbits about another body.

scientific method – the process by which scientific knowledge is acquired and refined.

sedimentary rock – rock formed by deposition of small particles from suspension (or solution) in water, followed by compaction over time.

selection effect – the bias of a particular conclusion that results from using a non-representative sample of observational data.

SERENDIP – the Search for Extraterrestrial Radio Emissions from Nearby Developed Intelligent Populations, a program run by the University of California at Berkeley using equipment mounted piggyback on the Arecibo radio telescope in Puerto Rico.

SETI – the Search for ExtraTerrestrial Intelligence, in the form of radio signals from space.

sexual reproduction – reproduction by the union of two different types of cells (egg and sperm), which combine their DNA to form a daughter cell different from either parent.

solar system – the Sun and its planetary system; it includes the planets, dwarf planets, asteroids, comets, moons, and other materials that are gravitationally bound to the Sun. (Occasionally 'solar system' is used as a generic term for a planetary system about another star.)

solar wind – an outrushing of charged particles – mostly protons and electrons – from the Sun's upper atmosphere.

special relativity – the theory that explains the measurement of objects and events in systems moving at relativistic speeds (close to the speed of light).

spectral type – a letter designation used to classify a star according to the appearance of its spectrum, which is primarily dependent on the star's temperature; in order from hot to cool, the principal spectral types are O, B, A, F, G, K, and M.

spectrum – different types of radiation, distinguished by their different wavelengths (or frequencies or energies).

speed – the rate of change of the position of a wave crest or a particle as it moves through space.

spiral arms – the luminous spiral patterns observed in the disks of spiral galaxies.

spiral galaxy – a galaxy in which most of the stars orbit within a flattened disk, which may exhibit spiral patterns radiating from a nuclear bulge at the center.

star – a self-luminous sphere of hot gas, normally powered by nuclear fusion reactions.

stasis – relatively long time periods of species stability, bounded by bursts of evolution, as described in the theory of punctuated equilibrium.

strong nuclear force – a very strong, short-range force that binds protons and neutrons together in the nucleus.

subduction zone – a region where tectonic plates collide, with one of them being driven underneath the other, back into the mantle to be re-melted.

sugar – one of the simpler forms of carbohydrates, such as the monosaccharide, glucose ($C_6H_{12}O_6$) or the disaccharide, sucrose ($C_{12}H_{22}O_{11}$).

supergiant – one of a group of stars located across the top of the Hertzsprung-Russell diagram; they are generally larger and cooler than their main sequence predecessors (stars of the highest mass).

supernova – the explosion of a massive star at the end of its life, following the collapse of its iron core.

supernova remnant – the luminous gaseous nebula formed by ejection of the outer layers of a massive star during a supernova.

supervolcano – a volcano that erupts periodically, on much longer time scales (a few hundred thousand years) and with much more energy than a regular volcano; an example is the Yellowstone supervolcano, which erupted about 640,000 years ago and appears to have a similar time interval between eruptions.

surface gravity – the acceleration of gravity at the surface of a planet, moon, star, etc.

T

terrestrial planets – small, dense, rocky (Earth-like) planets.

theory – a well-established hypothesis that provides very good explanations for the observational evidence.

tidal locking – synchronous rotation and revolution of an orbiting body, in which the rotational period and the orbital period form a simple whole number ratio (1:1, 2:1, 3:2, etc.); caused by gravitational tidal forces of the central body acting on (and restricting the rotation of) the orbiting body.

time dilation – the increase in the duration of an event in a moving system as measured by an observer in a different reference frame.

transit – the passage of a small body in front of a much larger body, such as a transit of Venus across the Sun's disk.

tsunami – a huge ocean wave (often called a tidal wave) produced by motion of the Earth's crustal plates or by impact of a meteoroid.

Tunguska event – the explosion of a comet or an asteroid over the Tunguska region of Siberia in 1908, which caused considerable devastation of a sparsely populated forested area.

U

unidentified flying object (UFO) – an object or light observed in the sky, the nature of which is not immediately obvious to the observer.

universe – everything there is, which appears to include billions of galaxies.

V

Van Allen radiation belts – two toroidal regions around the Earth that contain charged, radiating particles trapped by the Earth's magnetic field.

volcanic outgassing – the process by which gases are released from the interior of a planet through volcanic eruptions.

W

water hole – a relatively noise-free range of frequencies in the microwave region of the spectrum, bounded by the natural frequencies of the neutral hydrogen atom (H) at 1420 MHz and the hydroxyl radical (OH) at $\approx$1660 MHz; the fact that these two species combine to form water (H_2O) – an important molecule for terrestrial life – leads some to speculate that aliens may attempt to communicate using this range of frequencies.

wavelength – the distance between two adjacent crests of a wave.

white dwarf – one of a group of stars located in the lower left corner of the Hertzsprung-Russell diagram; they are generally very small and relatively hot, and do not sustain nuclear reactions in their cores.

Index

LaVergne, TN USA
11 December 2009
166651LV00004B/1/P